Flexibility in Engineering Design

FLEXIBILITY IN ENGINEERING DESIGN

Richard de Neufville and Stefan Scholtes

The MIT Press
Cambridge, Massachusetts
London, England

For information about special quantity discounts, please email special_sales@mitpress .mit.edu

This book was set in Syntax and Times Roman by Toppan Best-set Premedia Limited. Printed and bound in the United States of America.

Library of Congress Cataloging-in-Publication Data

De Neufville, Richard, 1939–
Flexibility in engineering design / Richard de Neufville and Stefan Scholtes.
    p.   cm.—(Engineering systems)
Includes bibliographical references and index.
ISBN 978-0-262-01623-0 (hardcover : alk. paper)
1. Engineering design. 2. Modularity (Engineering) 3. Engineering economy. 4. Flexible manufacturing systems. 5. Manufacturing industries—Risk management. I. Scholtes, Stefan. II. Title.
TA174.N496   2011
620′.0042—dc22

                                                                                    2011002055

10 9 8 7 6 5 4 3 2 1

To
Ginger, Julie, and Robert
and
Ingrid, Lukas, Philipp, and Alexander

Contents

Series Foreword

Engineering Systems is an emerging field at the intersection of engineering, management, and the social sciences. Designing complex technological systems requires not only traditional engineering skills, but also knowledge of public policy issues and awareness of societal norms and preferences. In order to meet the challenges of rapid technological change and of scaling systems in size, scope, and complexity, Engineering Systems promotes the development of new approaches, frameworks, and theories to analyze, design, deploy, and manage these systems.

This new academic field seeks to expand the set of problems addressed by engineers, and draws on work in the following fields as well as others:

- Technology and Policy
- Systems Engineering
- System and Decision Analysis, Operations Research
- Engineering Management, Innovation, Entrepreneurship
- Manufacturing, Product Development, Industrial Engineering

The Engineering Systems Series will reflect the dynamism of this emerging field and is intended to provide a unique and effective venue for publication of textbooks and scholarly works that push forward research and education in Engineering Systems.

Acknowledgments

The Government of Portugal provided extensive support for the preparation of this book through its sponsorship of the MIT-Portugal Program, a major collaborative effort to strengthen university research and education in Engineering Systems analysis and design. We are also grateful to BP, Laing O'Rourke, the MITRE Corporation, and the MIT-Cambridge University Alliance for their encouragement and continued interest in our work.

Many colleagues have encouraged and collaborated in our efforts to develop and demonstrate the value of flexibility in design. Notable among these are David Geltner, Manuel Heitor, Paddy O'Rourke, Bob Robinson, Nicos Savva, and Olivier de Weck. We have also appreciated and benefited from the network of colleagues who have critically reviewed and shaped our work. These include Luis Abadie, Gregory Baecher, Chris Caplice, José Chamorro, João Claro, Gail Dahlstrom, Marla Engel, Johannes Falck, Nuno Gil, Michael Haigh, Paulien Herder, Qi Hommes, Christopher Jablonowski, Houyuan Jiang, Vassilios Kazakides, Afonso Lopes, Ali Mostashari, Robert Pearce, Danny Ralph, Danielle Rinsler, Sam Savage, Joaquim da Silva, Eun Suk Suh, Joseph Sussman, and Angela Watson. The challenging support of doctoral and postdoctoral students has been invaluable. Our thanks go especially to Jason Bartolomei, Michel-Alexandre Cardin, Markus Harder, Rania Hassan, Rhonda Jordan, Konstantinos Kalligeros, Yun Shin Lee, Jijun Lin, Niyazi Taneri, Katherine Steel, Tao Wang, and Yingxia Yang.

Introduction

This book focuses on the challenge of creating best value in large-scale, long-lasting projects. It does this by directly confronting the central problem of design: the difficulty in knowing what to build, at what time. Indeed, to get the best value, we need to have the right facilities in place, when we need them. However, we cannot know what will happen in the future. No matter how hard we try to predict long-term requirements, the forecast is "always wrong." Trends change, surprises occur.

To achieve the best results, we need to adapt to circumstances as they arise. We need to have designs that we can modify easily to take advantage of new opportunities—or to mitigate adversities. The future is uncertain. Design that does not account for a range of possibilities that may occur over a long lifetime runs the risk of leaving significant value untapped—or incurring major losses. An uncertain future provides a range of opportunities and risks. We can deal best with these eventualities and maximize our expected value if we build flexibility into design.

This book helps developers of major projects create value by using the power of design flexibility to exploit uncertainties in technological systems. We can increase the expected value of our projects significantly by designing them cleverly to deal with future eventualities. Flexible design greatly increases our opportunities for success, as this book illustrates throughout. Designs we can adapt to new circumstances enable us to avoid downside risks and exploit opportunities. We can use flexible design to improve our ability to manage financial and social risks and opportunities. Technical professionals who can plan and execute a project to adapt to new circumstances can substantially increase the value obtained.

This book is for all current and future leaders of the development, operation, and use of large-scale, long-lasting engineering systems. Your current or prospective responsibilities may include, but are not limited to, projects implementing:

• *Communication networks* Fiber-optic cables, cellular devices, and fleets of satellites;

• *Energy production, transmission, and distribution* Thermal and nuclear generators, hydroelectric plants, wind farms, and other renewable energy sources;

• *Manufacturing* The production of aircraft, automobiles, computers, and other products;

• *Real estate* Residential and commercial high-rise buildings, hospitals, and schools;

• *Resource extraction* Oil exploitation and refining, mining, and smelting;

• *Transport* Airports, highways, metro lines, high-speed rail, ports, and supply chains; and

• *Defense systems* Aircraft, ships, and armaments of all kinds.

The common feature of these long-lasting engineering projects is that they are all subject to great uncertainties. It is impossible to know circumstances and needs 10, 20, or more years ahead. Moreover, technology changes rapidly and disrupts previous assumptions and forecasts. New technologies both create new opportunities and make previous investments obsolete.

This book is for the entire project team, including current or prospective:

• *Designers* The engineers and architects who create the physical implementations;

• *Financial analysts* The estimators of the value of different designs and so shape them;

• *Clients* The owners, public officials, and program managers accountable for the projects;

• *Investors and lenders* The shareholders, banks, pension funds, and others providing the capital for the investments;

• *Managers* The controllers of the facilities as they evolve over their useful life;

• *Users* The operators over the system, such as airlines benefiting from air traffic control facilities, or the medical staff of a hospital; and

• *Regulators* The authorities responsible for safeguarding the public interest in these projects.

Members of the project team all share the common challenge of creating and implementing flexibility in design. To succeed, they need to work together over time. A clever design that can adapt to new opportunities will prove fruitless unless the system managers understand the design and can organize to use it. Conversely, the best system managers may have little scope to cope with unforeseen circumstances if the designers have not configured the project with the flexibility to adapt. Thus, even though team members may participate in the development and operation of the system at different times, and may not deal with each other directly, they all need to work together to achieve the best results for the system.

Flexibility in design maximizes the expected value of a system over time. It enables owners and operators to adapt the system for optimal performance as its requirements and opportunities evolve over its useful life. Project team members will thus be most effective when they work together to integrate planning, design, and management activities from conception to eventual shutdown of the project. Creating best value in large-scale, long-lasting projects requires a sustained team effort. Success involves more than applying special techniques; it entails a way of thinking about systems and implementing them. Flexibility is a fundamental approach to systems design.

Organization of This Book

This book is intended to suit a range of readers interested in using flexibility to improve the value of complex engineering systems. It has three parts.

Part I provides a high-level overview of the concepts and methods of flexibility, why it is necessary, and how it delivers value. It gives sufficient information to senior leaders who want to understand the general issues around flexibility. It also motivates the later chapters, which examine elements of the overview in greater detail.

Part II presents the methods needed to identify, select, and implement the kinds of flexibility that provide the best value. This section is for designers and analysts who need to justify and implement flexible design. It covers the range of necessary techniques: procedures to forecast and anticipate a range of uncertainties; methods to identify the most promising kinds of flexibility to use; tools for evaluating and choosing the best flexible designs; and ways to implement flexible designs successfully over the life of the project.

The appendices in the last part of the book provide more detailed supporting explanations of the analytic tools and concepts used to identify and justify flexibility in design. Readers may benefit from one or more of these appendices depending on their interests and needs. The appendices present brief but comprehensive presentations of the mechanics of economic evaluation and discounted cash flows, the economic rationale for phased development, the mechanics of statistical analysis used in forecasting, the process of Monte Carlo simulation to explore complex scenarios, and the basic financial concepts of options analysis. Importantly, they provide a detailed discussion of the "flaw of averages," the conceptual pitfall that traps so many designs in underperformance.

HIGH-LEVEL OVERVIEW

1 Flexible Design: Its Need and Value

We don't even know what skills may be needed in the years ahead. That is why we must train our young people in the fundamental fields of knowledge, and equip them to understand and cope with change. That is why we must give them the critical qualities of mind and durable qualities of character that will serve them in circumstances we cannot now even predict.
—John Gardner (1984)

The Future Is Uncertain

Technological systems can quickly become obsolete. New developments continually arise to displace established technologies. What was state-of-the-art yesterday may be out of date tomorrow. We see this in our everyday lives. Consider the distribution of music, for example: In a few decades, it has gone from vinyl records, to tapes, to CDs, to downloading tunes wirelessly onto miniature portable devices.

What happens to consumers also happens to large industries. The recent development of global communications offers several examples of unexpected rapid change. Much to the surprise of their developers, the Iridium and Globalstar satellite telephone systems were obsolete the moment they came into being—ground-based cell phones had become universal (see box 1.1). The examples continue: Wireless is substituting for landlines; satellite broadcasting is eliminating the need for local stations. Disruptive technologies pervade our lives.

Unexpected changes can create both gains and losses. System designers often equate uncertainties with risks—and therefore with bad things. However, uncertainties can also create new opportunities. As the Internet has shown us, unexpected changes can create benefits that the original developers did not imagine. The future is as much about opportunities as risks. The examples in box 1.1 indicate that when thinking about uncertainties, we should not simply worry about downside risks—we need to keep upside potential in mind.

Box 1.1
Technological surprises

The Iridium Fleet of Communication Satellites

This case illustrates the sensitivity of technological projects to rapid changes in context. The Iridium fleet of communication satellites was a superb technical development—but a miserable financial failure.

Iridium originally consisted of more than 60 satellites that communicated with one another and any point on earth. It provided consumers with wireless telephone service from any location to any other, provided they took the three-pound satellite phone outdoors. Motorola designed Iridium in the late 1980s and deployed it a decade later. By that time, it was commercially obsolete—cell phone technology had swept the market.

Iridium went bankrupt and sold for $25 million, about 1/2 percent of the $4 billion investment.[1]

Global Positioning System (GPS)

The U.S. military originally developed GPS to control long-range missiles. The heart of GPS is a fleet of satellites constantly beaming signals, like lighthouses in the sky. Receivers can automatically triangulate these beams to locate themselves very precisely. Such chips are now commonplace in civilian applications. For example, aircraft can position themselves accurately when no radar is available. Cell phones have GPS. Drivers and hikers use GPS to find their way in remote areas.

GPS has created tremendous opportunities and value in ways unsuspected by the original designers. Because they did not anticipate this tremendous commercial success, they did not build the original GPS with any capability to benefit from it—they did not incorporate any way to charge a fee for the service.

New technology affects the value of investments directly and indirectly because of the way it changes patterns of demand. Advances may have complicated, unanticipated ripple effects. Improved health care, for example, has increased life expectancy, which in turn has contributed to a greater population of older patients with chronic diseases and complex co-morbidities. In general, the ultimate impacts of technological developments are complex and uncertain.

The potential benefits of any venture also depend on the vagaries of markets and many other factors. A copper mine may be lucrative if the price of copper is high but not worthwhile if demand changes and prices drop. The benefits of any process also depend on its productivity; the skill, experience, and commitment of staff; the success of marketing campaigns; the speed of diffusion of use; and many other factors.

As this chapter shows, the bottom line is that we cannot count on accurately forecasting the long-term benefits and costs of technological systems. In general, the future value of these investments is highly uncertain. This is the reality that confronts designers, analysts, clients, investors, managers, users, and regulators.

Standard Methods Are Inadequate

Unfortunately, design methods do not deal with the reality of rapid change. Standard practice proceeds from a set of deterministic objectives and constraints that define what designers must accomplish. These mandates go by various names: Systems engineers think of them as "requirements," architects refer to "programs," and property developers and others think in terms of "master plans." By whatever name, these restrictions channel designers toward a fixed, static view of the problem. In the case of the Iridium communications satellites, for example, the designers sized the fleet for worldwide use assuming 1 million customers in the first year of operation—they made no provision for the possibility of far fewer customers or a narrower service area. Likewise, in the extractive industries, it is usual to base design on an assumed long-term price of the commodity despite the fact that the prices of raw materials fluctuate widely. In practice, we "design to specification" when we should "design for variation."

In the same way, standard procedures for selecting designs generally do not deal with the possibility of change. The standard methods for ranking possible choices refer to the "cash flow" of an investment, that is, to the stream of benefits and costs that would occur in each period of the project if the conditions assumed were to exist. In practice, the evaluation process usually discounts this unique flow and brings it back to a reference time to create measures such as the net present value (NPV), the internal rate of return (IRR), or the benefit/cost ratio (see appendix B for details). None of these approaches recognizes two routine features of large projects:

• The assumed conditions, such as demand and price, constantly change; and

• Management might—and it generally does—eventually decide to change the system in response to new circumstances.

Therefore, the initial business case analysis used to select design solutions often falls apart later on in the project. Consequently, the path chosen for the project may be less than optimal.

The Standard Methods Are Passive

Standard methods do routinely explore how designs might react to new circumstances and how these might change future benefits and costs. Analysts calculate how various important factors—prices, market share, and rate of innovation—affect the cash flows and overall value of the projects configured to satisfy stated requirements.

The standard process designs projects based on a limited set of assumptions and then considers uncertainties. The focus is on creating robust designs that will perform satisfactorily under various uncertainties and possible stresses. This reflects a bunker mentality: Will we be able to survive adverse futures? Will we be able to sustain risks?

The standard process does not design with the uncertainties in mind. System designers do not generally explore how changes in specifications and market factors might change the design itself. The examples in box 1.2 illustrate what happens. In short, the usual design and evaluation procedures focus on an unrealistically narrow description of the possibilities.

Box 1.2
Standard design based on fixed assumptions: Oil platforms

Deep-water platforms for extracting oil and gas from sub-sea reservoirs, as in the Gulf of Mexico or offshore of Angola, can cost several billion dollars each.

The usual practice is to mandate designers to choose projects using a fixed price of oil, even though it varies widely and events have proven trends to be unreliable, as figures 1.1 and 1.2 show.[2] (The price assumed is a closely guarded corporate secret because of its importance in contract negotiations. It represents assumptions about longer-term prices and thus differs from immediate spot prices. It also varies by company and over time.)

The effect of assuming a fixed price is to ignore oil fields that would be profitable at higher prices. This means that when prices are high and exploiting secondary fields might be worthwhile, the platforms do not have easy access to these valuable reservoirs. Their exploitation would require a completely new project, which may not be economically feasible. The owners thus miss opportunities that a flexible, easily adjustable design would exploit.

Box 1.2
(continued)

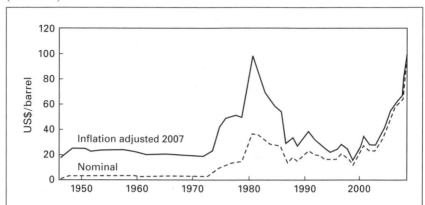

Figure 1.1
Historical prices of crude oil, both in nominal and constant value dollars. Notice
how the "trends" have frequently changed direction substantially: constant prices
until 1973, a sharp run-up over the next decade, followed by a decade of falling
prices and the more recent reversal.
Source: US Energy Information Agency, 2010.

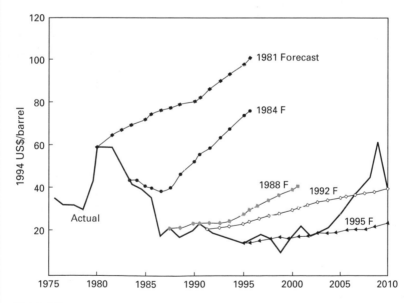

Figure 1.2
Forecasts of oil prices compared with actual prices. Notice how expert estimates
failed to anticipate reality, even in the short run.
Source: U.S. Department of Energy, compiled by M. Lynch.

The Standard Process Ignores Intelligent Management

The standard sensitivity analysis of a fixed design to alternate scenarios does not go far enough. It leaves out a crucial reality: The owners and operators of a project are intelligent. They will alter the design to suit the situations that actually develop, they may cut their losses by exiting from a project, and they may increase their profits by taking advantage of new opportunities. They will in any case respond actively to new circumstances rather than submitting to them passively, as standard evaluation procedures assume.

There is thus a great mismatch between what actually happens to a project over the span of its existence, as the system owners attempt to maximize its value, and standard ways of valuing and choosing between alternative designs. The reality is that future benefits and costs are uncertain, future adaptation to evolving circumstances is commonplace, and designers face a broad range of possible circumstances. Yet the standard valuation procedures assume that designers can design adequately around a single concept of future benefits and costs—and that users will not deviate from such a plan.

This mismatch matters. In general, there is a great difference between the value associated with the reality of many possible futures and the value calculated on the assumption of a single future. This gap exists because equal up-or-down variations in the imagined futures do not translate into equal variations in values. All the possible futures do not average out happily to the most likely cash flow. It is a fundamental mistake to assume that the average project value can be calculated on the basis of cash flow derived from average project conditions: This is when we fall victim to the "flaw of averages" (see chapter 2 and appendix A). It is imperative for designers to deal with the broad range of possible futures. Failure to recognize this leads to incorrect valuations of proposed projects and erroneous project selection. The design that appears best under standard analysis may be second rate.

Intelligent Management Anticipates Many Futures

Once we recognize that the future is uncertain, the intelligent thing to do is to prepare for its various possibilities. For example, if it looks like rain when we go out, the commonsense approach is to take an umbrella. "Be prepared" applies not only to our personal life but also to our professional practice.

Intelligent management anticipates and plans for a range of possible futures. It recognizes that it will react to circumstances as they change and so will prepare the way for this reality. Rather than react passively to what may come, it facilitates proactively the possibility of effective, timely responses to eventualities. It recognizes the possibility of risks and prepares exit strategies or other forms of insurance against their consequences. It also recognizes that advantageous opportunities may arise and anticipates ways of enabling the system to seize these benefits easily.

Intelligent Management Needs Design Flexibility

Flexible designs enable systems owners and managers to respond easily and cost-effectively to changing circumstances. Flexible designs come in many forms, each enabling different kinds of responses. In general, the system should incorporate several kinds of flexibility to protect against the range of different hazards and to exploit any opportunities that develop.

For example, the design of a car routinely incorporates many different flexible elements. It will have some kind of spare tire and jack in case there is a flat. It will have air bags to protect the occupants in case of a collision. It may have collapsible or removable seats in case the owner needs to carry oversize equipment. It might have a trailer hitch to enable users to attach things they might want to tow. In short, the modern automobile incorporates a range of flexibilities, enabling owners to mitigate downside risks and take advantage of upside opportunities.

Flexible designs fall into three major categories: those that enable the system to change its size, those that enable changes in function or capability, and those that protect against particular failures or accidents. For example:

• *Changes in size*   A design might be modular to permit the easy addition of capacity. A modular design might also facilitate contraction in capacity; for example, closing areas of an airport terminal during off-peak seasons when the capacity is not needed.

• *Changes in function*   The system might permit users to remove or add function. The design of a computer system is flexible with regard to both its hardware and software. Its USB ports enable users to upgrade from simple printers to multifunction devices, including scanners and fax machines. Its open software enables users to install voice over the Internet and transform their machine into a telephone.

• *Protection against accidents*   Systems normally feature a range of ways to protect against risks. These include protective systems, such as seat belts and airbags in cars. Redundant devices to back up a possible failure of key components are routine, such as second (and third) computers on aircraft.

In general, flexible designs include features that enable the system to respond to a range of possible circumstances, either automatically or under the direction of system managers.

Flexibility in Design Increases Expected Value

As both theory and many case studies demonstrate, flexible designs that are "prepared" for future possibilities can add great overall value to a project. Flexible designs enable owners, managers, or operators to adjust the design to new circumstances. When the future is unfavorable, they can avoid bad consequences. When the future offers new opportunities, design flexibility will enable them to take advantage and benefit from those possibilities (see box 1.3).

Flexibility in design can easily lead to significant improvements in overall expected benefits. The case studies we report in this book show increases of up to 80 percent in expected value. Flexibility provides a two-fold advantage: it limits possible losses and increases possible gains. Both actions increase the overall average value of a project. Even relatively small opportunities to make major gains or avoid disastrous losses can cumulate to important gains on average. For major projects costing billions, the combined value of flexibility in design can be worth hundreds of millions (see box 1.4).

Because we necessarily have to deal with an uncertain future, the value of any project is not a fixed number but an expectation over a range of possible futures. We can think of it as an average value over a range of good and bad outcomes. For this reason, design flexibility does not provide the best design under all circumstances. It would be cheaper, for example, to have a car without airbags and a spare tire—and this would be the better solution for the driver who never had an accident or a flat tire. Flexibility in design aims to provide improved solutions overall.

Sometimes a system will not use its flexibility, and its cost might be considered a waste. As for insurance, the value of flexibility has to be

Box 1.3
Flexible design: Tagus River Bridge

The Ponte de 25 Abril, the first bridge over the Tagus River at Lisbon, offers a good example of flexibility in design. The Salazar dictatorship inaugurated it in 1966 with a single deck for automobile traffic but with the strength to add a second deck at some future time. Moreover, they also built a railroad station under the toll plaza to minimize disruption in case Portugal ever decided to build rail connections.

A generation later, Portugal was a democratic member of the European Union, which allocated funds to develop commuter rail services throughout the region, and in 1999 the bridge received a second deck that carried these lines.[3]

When first built, designers of the bridge recognized that its ultimate capacity could be larger. Instead of trying to anticipate specific future requirements, they built for immediate use, with the flexibility to develop in many ways. Even if the original designers had tried to define future requirements, they could hardly have imagined the overthrow of the dictatorship and the development of the European Union.

The flexible design of the bridge saved money by not building too early or building unnecessary highway capacity. It also enabled Portugal to take advantage of the support of the European Union to extend rail traffic across the river.

judged in terms of its contributions over all possible futures. Both insurance and flexibility are justified by the value they bring when relevant events occur, not by their continual use. At the right price, we happily buy life and accident insurance every year and never complain about our failure to claim on these policies. It is the same with flexibility: Its value lies in helping us avoid bad situations or enabling us to benefit from opportunities in the right circumstances, not in whether we use it. Taking the proper overall perspective, flexible designs provide very significant opportunities for major increases in expected value.

What This Book Does

The book gives the leaders of major projects what they need to know to create value in technological systems by using the power of flexibility to deal with uncertainties and take advantage of them. It shows how project leaders can:

• *Recognize uncertainty* by replacing usual point forecasts with projections of realistic ranges of possible future outcomes;

Box 1.4
Flexibility leads to major gains: Satellite fleet

A detailed analysis of alternative ways to deploy geostationary satellites over different regions showed that a flexible system design, which enabled system operators to reposition satellites as demand for broadcast services changed, greatly outperformed the system "optimized" for the specified "most likely" pattern of demand.[4]

As table 1.1 shows, flexible design increases overall expected value. Instead of launching the final fleet right away, systems operators initially launch a smaller fleet, reducing initial capital expenditure, and therefore the amount at risk and potential losses. The flexible design, however, allows the capture of the upside, too, when operators deploy the second module, sized and located according to actual need, thereby obtaining a maximum value if demand exceeds initial capacity.

Table 1.1
Comparison of value of "optimized" and flexible designs for a satellite fleet

Design	Present value, $ millions			
	Expected	Maximum	Minimum	Fixed cost
"Optimized"	49.9	192	−162	−393
Flexible	95.8	193	68	−275
Which better?	Flexible	Flexible	Flexible	Flexible

The design "optimized" for a single forecast performs poorly on average across the range of possible scenarios.
Source: Hassan et al. (2005)

• *Identify desirable kinds of flexibility* that will enable the system to deal with the kinds of uncertainties it faces;

• *Understand and communicate the way flexibility adds value* to design;

• *Estimate the specific value that flexibility* contributes to their project; and

• *Implement a development strategy* that profitably exploits the advantages of flexibility.

In a nutshell, the book develops critical thinking about flexible design and provides a framework for presenting flexible designs.

The central message is that designing a system with the flexibility to adapt to future needs and opportunities greatly increases its long-term expected value, compared with standard traditional procedures for developing and implementing projects. In this book, we demonstrate this point with a wide range of practical applications.

Our book is pragmatic. It shows how project leaders of technological infrastructure can achieve extraordinary benefits, avoid future downside risks, and profit from upside opportunities by building flexibility into their designs. It provides a four-step process for developing design flexibility:

• *Step 1*   Recognize the major uncertainties the project or product is likely to encounter. This step identifies the kinds of situations where flexibility in the system might help.

• *Step 2*   Identify the specific parts of the system that provide the kind of flexibility best suited to deal with the uncertainties recognized in step 1.

• *Step 3*   Evaluate alternative flexible designs and incorporate the best into the design.

• *Step 4*   Plan for eventual implementation of the chosen flexibilities by both making arrangements with the stakeholders in the process and monitoring the conditions that would indicate whether and when to exercise the design flexibility and adapt the system to new circumstances.

Box 1.5 illustrates the process, which we describe in detail in part II.

Box 1.5
Application of the four-step process: High-rise building

Many developers have used "vertical flexibility" in the design of their buildings. The development of the Health Care Service Corporation building in Chicago (figure 1.3) illustrates the process.[5] The original design for this building had the strength, the space for elevator shafts and stairs, and the planning permissions to add 24 more stories to the original 30-story skyscraper built in the 1990s.

The four-step process proceeds as follows:

1. *The architects/developers recognize the client's uncertainty* about the amount of space needed, for example, because it is not possible to be sure about long-term growth or how zoning regulations might change to allow greater height.

2. *They identify that vertical flexibility is the only realistic possibility* because they cannot increase the ground area available for development.

Box 1.5
(continued)

3. *They explore numerous design alternatives*, involving different numbers of floors for the first and subsequent possible additions, estimate how each might perform under the range of future scenarios, and choose the arrangement that provides the best set of metrics overall.

4. *They plan for implementation* by making arrangements with the multiple stakeholders involved in the execution of the vertical flexibility and monitoring developments to determine when (and if) they should use their design flexibility. They obtain planning permission for expansion, operational support from the tenants, financial commitments from the bankers, and so on. They then keep track of their needs for additional space. The owners inaugurated the additional phase in 2010.

Figure 1.3
Vertical expansion of Health Care Service Corporation Building in Chicago in center of image: phase 1 (left) and phase 2 (right).
Source: Goettsch Partners release to Wittels and Pearson, 2008.

2 Recognition of Uncertainty

In the early 1980s the consultants McKinsey and Company were hired by ATT to forecast the growth in the mobile market until the end of the millennium. They projected a world market of 900,000. Today [in 1999] 900,000 handsets are sold every three days.
—A. Wooldridge (1999)

This chapter indicates how we might best anticipate the future for long-term technological systems. Modesty is the best policy. We need to recognize the limits to human foresight. We need to recognize that forecasts are "always wrong" and that our future is inevitably uncertain. We thus need to look at a wide range of possible futures and design our projects to deal effectively with these scenarios.

The success of a design lies in how well it provides good service to customers and the public over time. If we create efficient systems that continue to fulfill actual needs, in the right place and at the right time, we will be successful. If the systems do not provide the services needed conveniently and in good time, our creations will have little value. Successful design thus depends on anticipating what is needed, where and when. Success depends on the way we understand what the future might bring.

Unfortunately, it is impossible to predict exactly what the future will bring over the long term and over the life of systems. Experience demonstrates, again and again, that specific forecasts turn out to be far removed from what actually happens. Exceptionally, we might be able to look back after 10 or 20 years and see that our long-term forecast was accurate, but few of us are ever so lucky.

Moreover, because so much about the future is unpredictable, we cannot ever hope to achieve reliable long-term forecasts. We need to recognize the inevitable uncertainty of any prediction and give up on the impossible task of trying to develop accurate forecasts of long-term futures.

Instead, we need to adopt a new forecasting paradigm that focuses on understanding the range of circumstances that might occur. This distribution of possible futures is often asymmetric, with rigid limits on one side and almost none on the other. For example, the price for a commodity is not going to drop below zero, but it may become very high. Similarly, the maximum demand a plant can serve might be limited by its capacity, yet the minimum could be zero. We need to engage with ranges of circumstances and their probabilities to appreciate the context in which our products and systems will function, to be aware of the risks that may threaten them and the opportunities that might occur. This focus on the likely range of possibilities is fundamentally different than the standard approach to forecasting, which looks for the right or best single number.

Uncertainty Matters

Forecast uncertainty matters. If we do not take a range of possible outcomes into account from the start, we are almost certain to get misleading results from our economic appraisals and to select projects and designs incorrectly. Why is this? The intuitively obvious thought is that we could base our designs and evaluations on the most likely or average forecasts and let the underpredictions balance out the overpredictions. Why is this not right? These are good and important questions. The answer in a nutshell is that this intuition misses an important point. If we base our work on average forecasts, we fall into the trap of the "flaw of averages" already mentioned in chapter 1. When we do the mathematics correctly, we find that basing designs and evaluations of projects on average is almost certainly wrong. To design and choose projects correctly, we absolutely need to consider the range and distribution of possible outcomes.

The "flaw of averages" justifies the concern with the distribution of possible outcomes. Understanding the difficulties associated with this fallacy is central to dealing properly with risks and uncertainties. We therefore start this chapter with a section that develops the understanding of the "flaw of averages."

The bottom line is that we need to develop our understanding of the range of possible futures. In this way, we can anticipate what might be needed and provide for these possibilities from the start. This approach enables us to adapt our designs to what we eventually discover is actually needed or desirable and consequently increase their value—which is the point of this book.

The Flaw of Averages[1]

The "flaw of averages" refers to the concept that it is not correct to calculate the average value of a project based on its performance under average conditions. This may seem peculiar. Why don't "average inputs" lead to the "average outcome"? Here we explain the basic elements of this puzzle.

We start by noting that the "average outcome" of a project or product is the average of its performance under all the possible scenarios in which it might exist. This is what we need to know. We then need to recognize that different scenarios can have quite different consequences for a project. Higher demand for a product may—or may not—compensate for a similar amount of lower demand. Higher demand may not increase revenues, for example, if the system does not have the capacity to meet it. Conversely, lower demand could be disastrous if the resulting low revenues do not cover the cost of the system and lead to bankruptcy. The asymmetry between the results associated with differences in scenarios can be crucial. We cannot appreciate this by looking at the average scenario. We can only correctly evaluate the real results of a system by looking at a range of scenarios.

Let's take an example. Imagine a game in which you roll a die and will receive $3 if the resulting number is 3 or more—and nothing otherwise. Assuming that all the possibilities from 1 to 6 are equally likely, the average number showing will be 3.5. If you calculate your payoff based on this average number, you obtain a payoff of $3. However, the average payoff is only $2! This is because there is a 1 in 3 chance of rolling a 1 or a 2, which leads to a payoff of $0 instead of the $3 calculated based on the average of the numbers on the die. This loss in the "downside scenario" is not counterbalanced by additional gains in the equally likely "upside scenario" of rolling a 5 or a 6, in which case you still only get $3. You lose in the downside yet do not gain more in the upside scenario.

This example is widely replicated in industry whenever it has to plan production capacity. Auto manufacturers, for instance, typically design plant sizes based on their most likely forecasts of sales of new models. If the model is a success and demand is greater than expected, they can only expand capacity to a limited degree (and substantial cost) by working overtime or extra shifts. They will not be able to capture fully the benefits of the upside scenario. However, if demand is poor, they definitely face the full downside and may suffer large losses. (Box 2.1 provides a worked illustration of this, and we give full details about the issue in appendix A.)

The flaw of averages effect is easy to understand—yet planners neglect it remarkably often. Upsides and downsides of input uncertainties (numbers on the die, demand for cars or hotel beds) may balance out nicely around their average, but these upsides and downsides often have asymmetric effects on the measure of value for system performance (such as its profit or present value). We simply cannot expect system performance to balance out around the performance calculated on the basis of average scenarios.

The Need for Forecasts

Forecasts are fundamental to design. We plan and implement systems around anticipated future demands and opportunities. Our estimates of what we will need shape what we build, where we locate it, and when we implement it. These projections focus our thinking and determine major design characteristics.

Meanwhile, the value of any design derives from how well it matches the realizations of future demands and opportunities. The value of a system depends on the extent to which it matches ongoing needs, at the right time and place. A "bridge to nowhere" may be technically brilliant, structurally elegant, and efficiently and safely constructed. However, if it leads nowhere, serves little traffic, and does not fulfill a meaningful function, it has little value. Again, the development of the Iridium system to provide worldwide telephony illustrates the point. Iridium was an astounding technological achievement but a financial disaster. It did not provide good value for the money invested in it. Good design does the opposite of this: It provides good value in terms of fulfilling actual demands.

Forecasts determine the valuation of a design. Consequently, the usefulness of our estimates of value depends on how well our projections of the range of future needs and opportunities match those we eventually encounter. If we manage to anticipate possible scenarios correctly, our design has a good chance of being successful. If our estimates are misguided, then our design may fail—either through exposure to possible risks or by missing possible opportunities.

Box 2.1
Flaw of averages: Capacity constrained facility

A simple example illustrates the flaw of averages. Consider the valuation of an investment in hotels. Like most other infrastructures, the value extracted from these buildings depends crucially on demand, which is highly uncertain.

Consider a company planning to build a hotel to meet a projected annual demand of 750,000 overnight stays at $150 per night. It understands that there is demand risk. There might well be a 25 percent chance of a low-demand scenario of 500,000 overnight stays at a reduced price of $100 per night and an equally likely high demand scenario of 1,000,000 overnight stays at $200 per night, as table 2.1 indicates. The uncertainty in demand averages out to the projection.

Following standard practice, the company would value the opportunity according to the average condition. Deducting annual costs of $100 million, it anticipates annual revenues of $112.5 million and a net profit of $12.5 million.

What happens in the downside and upside scenarios? If the downside scenario occurs, then the company will only sell 500,000 overnight stays at a reduced price of $100. If the high-demand scenario occurs, the company will be able to charge the higher price of $200 but will still only be able to accommodate 750,000 overnight stays. The capacity constraint limits the gains. Therefore, the economic loss of the downside scenario is much larger than the gain of the equally likely upside. Specifically, considering a 25 percent chance for the upside and downside scenarios, the expected return is not $12.5 million but $6.25 million $[- (0.25 \times 150 + 0.5 \times 112.5 + 0.25 \times 50) - 100]$, half the estimate based on average project conditions.

Except in rare cases, an economic evaluation based on average values leads to an incorrect result. Because of constraints and other discontinuities, we can only obtain a correct valuation by considering the results associated with the distribution of scenarios.

Table 2.1
Summary of data for hotel example

Spaces			Money per year, $US millions		
Demand	Used	Daily rate	Revenues	Costs	Profits
1,000,000	750,000	200	150	100	50
750,000	750,000	150	112.5	100	12.5
500,000	500,000	100	50	100	−50

So the first question for project leaders and designers must be: How do we get the right kind of estimates of the future? There are two answers to this question. First, we must change our expectations about prediction. Current practice does not serve us well; it tends to develop forecasts that are unjustifiably precise. More importantly, long-term forecasts are reliably wrong. Second, as well as trying to anticipate likely futures, we must recognize the large uncertainties around these estimates. This is where modesty comes in: We must be modest about our ability to predict, and we must recognize uncertainty.

Standard Practice

Obviously, it is desirable to know the purpose of our design. Consequently, the development of large-scale, long-term systems and products always involves great efforts to forecast the future or, equally, to specify the requirements for future development. For commercial or civilian projects, such as the construction of oil platforms or the production of automobiles, these efforts are largely directed toward determining the context in which the system will perform—for example, the quantity of oil and gas in an oil field or the level of demand for types of vehicles. For military projects, the emphasis is on specifying the "requirements," that is, the specific functions that a future system must fulfill—which, of course, results from forecasts of what threats the system will have to face. Either way, traditional process directs the work toward defining future needs.

Point Forecasts

Standard practice leads to very precise predictions or specifications. The almost universal result from a forecasting process is a series of "point forecasts," precise numbers concerning what will happen in various years. The aviation forecasts developed by the U.S. Federal Aviation Administration provide a good example of this. Table 2.2 presents an extract from one of their forecasts. Note the precision—nine significant figures on a 20-year forecast.

Highly precise estimates of long-term futures are commonplace for all kinds of activities. For example:

• The International Energy Agency provides annual energy forecasts, indicating for example that "Demand for oil rises from 85 million barrels per day now to 106 mb/d in 2030"; and

Table 2.2
Forecast enplanements at major U.S. airports (those with FAA and contract towers)

Size of airport	Actual 2008	Forecast 2030
Large hubs	511,064,981	871,682,705
Medium hubs	154,655,693	225,541,181

Note assumed ability to forecast to a single passenger 20 years ahead.
Source: U.S. Federal Aviation Administration (2010), Table S-5.

• The U.S. Energy Information Administration predicted that "wind energy production in the United States will rise from 24.81 gigawatts in 2008 to 42.14 in 2030."

Such precise forecasts are not credible. The most skilled technicians, using sophisticated scientific instruments, operating in tightly controlled laboratories, would be fortunate to achieve such levels of accuracy in measuring known substances. Estimates of what might happen, in the messy real world, years ahead, cannot be anywhere near this good. Figure 1.2 shows how bad energy forecasts can be. If we are lucky, our predictions might come within a few percent of eventual reality—optimistically, the first two figures of the forecast might be correct.

Yet the common feature for almost all point forecasts is that they provide no indication of their uncertainty, let alone the degree of confidence we might have in them. They do not suggest the real possibility— the almost certainty—that actual situations may be very different than forecasts. Point forecasts are thus misleading. In fact, as we show below, there are typically large gaps between point forecasts and what actually occurs.

Specialized Professionals
Another important feature of standard practice is the way it is done. Specialized professionals usually manage the process of forecasting and typically work in separate agencies, companies, or divisions, distinct from the designers or planners for the systems and products under consideration. For example:

• Economists and statisticians develop models to predict the future levels of air traffic.

• Market analysts use consumer surveys and focus groups to predict future sales of automobiles.

• Geologists develop best estimates of the size of oil fields, which they hand over to the structural and other engineers designing oil platforms.

In short, a great professional distance can exist between the forecasters and those who use their numbers for design. Physical and institutional distance between these groups reinforces this disconnect.

This separation of tasks shifts responsibilities and provides convenient excuses for design professionals. If their product does not meet actual demands, they can feel absolved of responsibility. They can argue that they did what they were told and cannot be responsible for the forecasters' mistakes—while the latter are conveniently in some other organization and have probably left the scene by the time it is discovered that their long-ago forecast does not correspond to eventual outcomes.

The gap between the producers and users of forecasts is dysfunctional from the perspective of achieving good design. Each group of professionals works in its own world with its own criteria and expectations. The overall result is that designers use unrealistically precise estimates during the design process. Standard practice thus frequently becomes trapped into fixating on misguided concepts.

The professional context drives forecasters toward reporting precise results, even though it is easy for them to report ranges. From the outside, companies and agencies want precise forecasts and often ignore ranges when available. They routinely spend millions and millions on forecasting exercises. These fees impel analysts to provide detailed answers. Indeed, who would want to spend large amounts to get a report saying that we can only know the future market within plus or minus 50 percent of the forecast? From inside the profession, the tradition is to judge models according to statistical tests of how well their models of behavior match historical data—which are the immediate evidence at hand. The forecasting exercise is widely seen as a precise mathematical analysis subject to accurate calculation. Both the professional mind-set and the demands of managers buying the forecasts drive the forecasting process to produce detailed numbers and reassuring precision.

Complementarily, engineers want to work with specific numbers. Engineering is supposed to be analytic and precise, not fuzzy or ambiguous (see box 2.2). Furthermore, however complex the system, it is much easier to design to some exact requirement than to a range of possibilities. Moreover, managers want assurance that their plans will be on

Box 2.2
The gap between designers and forecasters

One of the chief designers of the Iridium satellite system for global tele-phone service visited MIT after the financial collapse of the venture. In sharing his experience with the design, he indicated that one of his greatest difficulties was obtaining a correct forecast of future use and by extension fixing the specifications of the system.

What the designer wanted was a definite fix on the number of users. He wanted to use this number to optimize the size and capabilities of the individual satellites and the fleet. The difficulty of getting this commitment from the marketing and forecasting team frustrated him.

He was, of course, asking the wrong question. No forecaster can antici-pate precisely the actual demand for an untested service that would be deployed a decade later in competition with other rapidly evolving forms of technology—cell phones in particular. In retrospect, the forecast used for the design was about 20 times too high.

The right question for the designer would have been: "What range of usage might exist?" The answer to this question would have given him some idea about how he could stage the deployment of the system, about actual demand, and about how to shape the future development of the system to accommodate actual needs.

target. Together, engineers and managers reinforce the demand for detailed, exact forecasts.

Unfortunately, the combined focus on precise forecasts is misplaced. The results are not worthwhile, as experience demonstrates. Sadly, it is hard to break the reinforcing professional dynamics that drive the focus on precise predictions. However, we are better off dealing with reality than wishes. In this spirit, we need to examine carefully what we can reasonably hope for.

The Standard Forecast Is "Always Wrong"

Forecasts for any year—about the future demand for products or services, future prices, future technological performance—rarely tell us what actually happens. The actual values or levels that exist in the forecast year are, as a rule, different than what most people predicted. After-the-fact comparisons routinely demonstrate a big gap between the forecast and the reality. It is in this sense that a good rule of thumb is: "The forecast is always wrong."[2]

Of course, by chance some forecasts turn out to be correct. When several forecasters make predictions, some projections will turn out to be closer to the true value than others, and some may even turn out to have been quite accurate. When the Swedish Central Bank examined 52,000 forecasts made by 250 institutions for inflation and economic growth in the United States, Japan, Great Britain, France, Italy, and Sweden during the period 1991–2000, they found that no organization was ever fully right—some would predict inflation well but not growth and vice versa.[3] In any year, some organization will turn out to be the most accurate, but this is largely a matter of luck—the winners change from year to year. There is no consistent ranking and no clear indication of which forecast will be right next time.

The general rule that forecasts for particular projects are "always wrong" is widely documented. Study after study has compared forecasts for a system to what actually happens. The results show substantial differences between what forecasters anticipated and the reality 5, 10, or 20 years later. Three examples illustrate such discrepancies.

Consider first the estimation of oil reserves. We might think that the amount of oil in a reservoir will be knowable. After all, the deposit came into existence millions of years ago and is fixed. However, the quantity of interest is not the overall size of the field; as for any mineral deposit, what we really want to know is the amount that our technology can extract from it economically at market prices. This amount depends on many unknown factors. For oil, the economical reserves depend on:

• Technology, which is constantly changing so that deposits once inaccessible are now available to us;

• The market price of the product—for example, the huge deposits of oil in the tar sands of Alberta can only be economically extracted if the price of oil is relatively high;

• The structure of the oil field—the extent to which it is fractured and transmits pressure that drives the oil and gas out;

• The quality of the oil—for example, its viscosity (which affects its flow) and its sulfur content (which affects its value); and, of course,

• The skill of the operators in managing the field by injecting water (to maintain pressure) and combining flow from different wells (to keep viscosity sufficiently low).

In short, forecasts of reserves must be full of uncertainties—as indeed they are (see box 2.3).

Box 2.3
Variation in the estimate of oil reserves

Figure 2.1 represents experience with two oil fields in the North Sea, normalized around the size of the original estimate.[6] The P_{50} curve represents development of the median estimate over time (i.e., at any given time, the geologists estimate that there is a 50 percent chance that at least P_{50} barrels of oil will be commercially worth extracting). P_{10} and P_{90} are the limits defining the range of the actual quantity expected to occur 80 percent of the time (the 80 percent confidence limits). Although we might imagine that the estimates would steadily converge on some value as more experience and information develops, it often happens that this experience reveals new finds or difficulties that trigger a "jump" in the range of estimates, as the graphs show.

In the case on the left, the range of the estimate does eventually narrow considerably, but the ultimate P_{50} estimate is less than half the original amount. In the case on the right, the range of estimates hardly narrows, and the P_{50} estimate nearly doubles. Notice moreover that over 1 year the best estimate rises dramatically, far exceeding what was considered the previous range of likely possibilities.

In this context, design teams often focus on the original best estimates— the left-hand side P_{50} values—to design the platforms and wells for the field. For the two fields shown in figure 2.1, this practice means that the resulting platform designs would be off by almost a factor of 2—about double the desirable size for the case on the left and only half what would be needed for the case on the right. Although the geologists know from experience that estimates can jump over time, and although operators constantly struggle with production uncertainties, standard design practice does not account for this uncertainty.

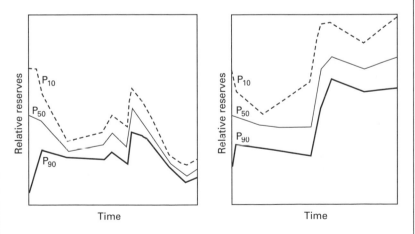

Figure 2.1
Variations in estimated oil reserves over time. In both cases the median P_{50} differed by a factor of two: it halved for the left-hand field, doubled the right-hand field.
Source: BP sources from Lin, 2009.

Consider next the matter of forecasting the cost of new systems. Collectively, the track record is poor. New high-technology systems are understandably uncertain. We might think, however, that we would be able to estimate costs reasonably accurately for long-established systems such as railroads, metro systems, and highways that we have been deploying for more than a century. But this is not the case. Exhaustive worldwide studies show that the cost estimates supporting design and investment decisions are routinely and significantly off. Figures 2.2 and 2.3 demonstrate this with reference to major highway and rail projects.[4] For road projects, planning estimates of costs were over 20 percent incorrect about half the time. For rail projects, the errors were even larger.

Apparent counterexamples of good predictions of future system costs are often not what they seem. The owners of Terminal 5 at London/ Heathrow airport boasted that they delivered the project on time and budget—a great accomplishment due to innovative management. What they did not advertise was that the project's final cost of more than £4 billion was well over twice the cost the developers originally anticipated when they decided to proceed with the project more than a decade before it was opened. As ever, the project changed, various issues arose, and the cost grew considerably. In any case, developers often meet their budget by cutting down on the design so that comparisons often do not compare like with like.

Consider finally the matter of forecasting demand for products or services. Demand is especially hard to anticipate correctly. It results from a complex interaction of market forces, competition from other providers or other services, and changeable personal preferences. When forecasting demand, we need to anticipate not only the overall, aggregate demand but also that of its components. For example, it is not good enough for an automobile manufacturer to have a good estimate of the total number of cars it will sell. It also needs to know how many different kinds of vehicles it could sell because it has to design and equip its factories differently to produce sedans, hybrids, minis, and sport utility vehicles. Any company or organization delivering a variety of products or services must be concerned with forecasting each of these elements, and this is especially hard to do (see box 2.4).

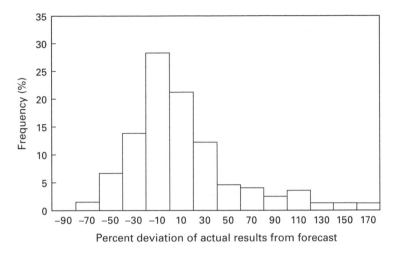

Figure 2.2
Discrepancies between the forecast and actual costs of road projects.
Source: adapted from Flyvbjerg et al., 2005.

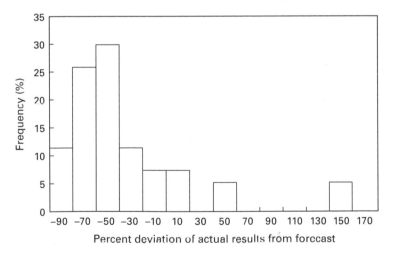

Figure 2.3
Discrepancies between the forecast and actual costs of rail projects.
Source: adapted from Flyvbjerg et al., 2005.

Box 2.4
Difficulty in predicting demand and its components

Forecasts for Boston airport illustrate the difficulty in predicting demand. In the case shown in table 2.3, the discrepancy between the overall forecast and reality was 10 percent over only 5 years. In retrospect, this was due to the burst of the dot.com economic bubble—one of those unexpected disruptive trend-breaking events that so frequently and routinely mess up forecasts.

The 10 percent discrepancy was not even across the board, however. The number of aircraft operations dropped by twice as much (23 percent), as airlines cut out flights by small regional aircraft (42 percent off the forecast) and squeezed more people into their big jets. This kind of differential change in demand over product categories is typical.

Table 2.3
Discrepancies between actual and 5-year forecast of emplaned passengers and flight operations at Boston/Logan airport

Category	Actual 1999	Forecast 2004	Actual 2004	% deviation
Passengers	13,532,000	14,500,000	13,074,190	10
Operations				
Total	494,816	529,129	409,066	23
Big jets	255,953	265,058	236,758	11
Regional	189,680	222,949	128,972	42
Cargo	11,267	13,180	12,100	7
General aviation	37,916	27,942	31,236	12

Source: Massport and U.S. Federal Aviation Administration.

Requirements Are Also Often Wrong

"Requirements" provide an alternative way to specify what a design or project should look like. Forecasts define a system's opportunities or demands—for example, the size of an oil field or the loads on a commuter rail system. They guide the designer to a set of functions the client wants the prospective system to fulfill. Requirements directly specify what these functions and capabilities should be.

Governmental and other organizations routinely focus on requirements for areas where numerical forecasts might not be available or useful. The military go to great lengths to specify the requirements for future systems, working with experts and consultants to define future threats, or the new capabilities their future strategies might require. Similarly, organizations in charge of air traffic control specify performance standards for their aircraft monitoring and controlling equipment.

Requirements frequently change. The things developers once proclaimed to be "required" often turn out to be no longer necessary. This happens when events or technology change the system managers' perceptions of what is needed. The evolution in the way the military and other agencies have thought about unmanned aerial vehicles (UAV) illustrates how this occurs (see box 2.5).

To sum up, estimates of what might be desirable and necessary are not reliable. Understanding this fact and incorporating it into design practice gives us the basis for creating the most realistic design for our products and systems.

Box 2.5
Changing requirements—unmanned aerial vehicles

Unmanned aerial vehicles (UAVs) are aircraft that do not carry a pilot.[7] They range from model planes operated by hobbyists to much larger aircraft with wingspans of about 20 meters and costing more than US\$8 million each. They differ from earlier drones in that ground operators stationed at great distances, often halfway around the world, can control them tightly—the U.S. Air Force directs flights over the Middle East from a base in Nevada.

UAVs became practical with the development of miniature electronics and global communications and are thus a product of the 1990s that represents a new technology. Users are gradually coming to understand how they can best use them.

The role of UAVs has changed dramatically as users have gained experience. Originally, the military saw them as surveillance aircraft that would loiter slowly over an area. The requirements were therefore for relatively slow vehicles (to make it easier to get good images) with modest payloads, such as cameras and electronic sensors. More recently, the requirements for the same vehicles have shifted toward carrying heavy loads (bombs weighing a quarter ton) and greater speeds (to escape anti-aircraft fire). Concurrently, civilian users, such as the Forest Service interested in spotting fires, have other requirements, such as those associated with sharing airspace with commercial and other manned aircraft.

Inescapable Obstacles to Accurate Forecasting

Forecasts are "always wrong" because there are two inescapable obstacles to good forecasting. The first consists of "trend-breakers." These are the ever-changing sets of surprises that disrupt our expectations. The second lies in the inevitable ambiguities that forever confuse our interpretation of historical records.

Trend-Breakers

These are events that disrupt the smooth continuation of what has been happening in the recent past. There are all kinds of trend-breakers:

• Economic crises, as in 1991, 2001, and 2008, and many other times;

• Political shifts, both peaceful (the creation of the North American Free Trade Agreement, for example) and otherwise (wars and revolutions);

• New technologies, such as the development of terrestrial cell phones that destroyed the potential market for satellite phones (Iridium, again);

• Discoveries, as illustrated in figure 2.1, which shows the doubling of the expected reserves in one oil field; and

• New market conditions, such as new environmental policies and regulations, the emergence of major new competitors, shifts in customer preferences, and so on.

Trend-breakers occur routinely. Although individual ones may come more or less as a surprise, records show that we must expect such surprises. They regularly disrupt long-term forecasts and create new conditions that designers and managers need to face. They may reshape large aspects of the international economy, as the burst of the housing and dot. com bubbles did in 2001 and 2008. They may have more limited regional effects, such as the opening of the Channel Tunnel, which redistributed traffic between continental Europe and the United Kingdom. They may affect an entire industrial sector, as mobile telephones did to communications. They may only disrupt a product line, as the development of light-emitting diodes (LEDs) is doing to the light bulb industry. Trend-breakers are real—and a basic reason why any trend must be associated with considerable uncertainty.

What Is the Trend?

Trends are ambiguous. Given the historical ups and downs, either in overall demand for some services or in our perceptions of the quantity of oil in a reservoir, how do we establish a trend? This is not an issue of mathematics or graphics. Indeed, the procedures for developing a trend from any given set of data are well established and routine. The issue is: What set of data should we consider for establishing a trend?

One important problem in establishing a trend is that the results depend on the number of years considered. If we choose a very short period, the short-term effects may dominate and obscure a long-term trend. For example, looking at the passenger traffic for Boston from 1999 to 2004 in table 2.3, we could see an average drop in traffic of about 0.7 percent a year. This is a direct result of the burst of the dot.com bubble and the concurrent 2001 terrorist attacks. However, choosing a very long period, such as 50 years, may cover several trends. Indeed, air traffic in the United States and Boston enjoyed a spurt of growth in the 1980s, following the national deregulation of airlines, the overall lowering of prices, and increased competition. That spurt has since dissipated, and we might best exclude it from any analysis of the current trends. Following that logic, the appropriate period for projecting might start somewhere in the 1990s. But when? Which data should we include to establish a trend?

The years selected for establishing a trend can change a long-term projection drastically because slight changes in the pointer multiply into substantial differences when projected far. The example in box 2.6 demonstrates how slight modifications in the range of historical data chosen

Box 2.6
Estimating trends

Figure 2.4 shows the evolution of the Singapore Property Price Index over 35 years. It has had its ups and downs, variously associated with the oil crisis around the early 1980s, the Asian crisis around 1995, the 2001 dot. com meltdown, and the financial crisis in 2008. Many other price indices reflect these or other fluctuations.

The trend we extract from such historical records greatly depends on the years selected, as table 2.4 illustrates. It shows that statistical estimates of the growth rate vary enormously. Moreover, there is no scientific reason to determine the best period for analysis. Analysts can make more or less reasonable arguments for longer or shorter periods to include specified periods or not. Ultimately, the choice is a matter of taste. The trends thus estimated are more or less arbitrary. Furthermore, estimates a few years apart may be very different indeed. Projections made in 2008 (before the crisis) differ greatly from those made 2 years later. Trends are not obvious!

Box 2.6
(continued)

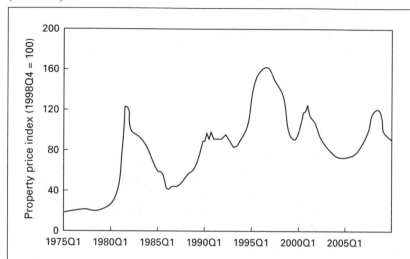

Figure 2.4
Evolution of Property Price Index for Singapore, 1975–2009.
Source: Zhang, 2010, based on data from the Singapore Urban Redevelopment Authority.

Table 2.4
Trends in the Singapore Property Price Index

Start of data series	Annual growth rate (percent)		Standard deviation (percent)	
	Include 2008–9	Exclude 2008–9	Include 2008–9	Exclude 2008–9
1975Q1	4.544	4.871	50.7	51.2
1986Q1	1.719	1.820	32.4	33.5
1993Q1	(2.332)	(3.343)	23.0	23.0
1999Q1	(0.021)	(2.260)	18.2	15.8
2004Q1	9.457	13.093	11.2	6.8

Estimates of historical growth rates and volatility are sensitive to arbitrary decisions made about the length of the period analyzed, as analysis the data on the property values in Singapore data illustrates.
Source: Zhang (2010), based on data from the Singapore Urban Redevelopment Authority.

can change the calculated trend enormously (in this case, from about −3 to +13 percent annual growth). Trends are not obvious: They are in the eye of the beholder.

What Drives the Trend?

What factors increase future development? What might encourage users to want to use new communication services? Or hybrid cars? Or any of the products and services our systems might deliver?

The process of determining the drivers of future forecasts is also highly ambiguous because it is rarely possible to separate out the factors that determine the change in anything in which we might be interested. As it turns out, most possible explanatory variables follow similar patterns, and it is almost impossible to unscramble their separate effects.

The pattern of change of most developments can often be reasonably described for a limited time by some annual rate of growth, call it r percent a year. This means that, starting from a current level L, the future level after T years is:

Future level, $L_T = L \,(1 + r)^T$

This pattern is exponential growth. Moore's Law describing the doubling of computer capacity every 2 years is one example of this phenomenon. Exponential changes can sometimes last a long time; Moore's description of change held for more than four decades from the 1960s. However, there are clearly limits to such patterns. In general, they last for a limited time. The sales of new products, for example, frequently grow exponentially during early stages and then slow down as markets become saturated.

Exponential change generally applies not only to the trend of the factor that we want to forecast—but also frequently to several other factors that might drive or otherwise influence the factor of interest. For example, the factors influencing the increased use of cell phones might be the steady drop in price for equivalent functionality, the increase in population, and changes in overall wealth—all of which might reasonably experience exponential change, at least for a while.

Because an exponential process describes the rate of change of many possible drivers, we cannot untangle their effects. Technically, potential drivers are likely to be highly correlated. The consequence of this high correlation is that it is not possible to have much—if any—statistical confidence in any of their effects. Box 2.7 describes examples of this

Box 2.7
Ambiguous analysis of causes of trends

A regular classroom exercise at MIT involves working with extensive historical data on traffic in Los Angeles. The purpose is to explore how the experts on the actual job arrived at their forecasts from these data. One phase considers the expert analysis that selected the relative importance of population and income as drivers of traffic growth based on the combinations that matched the historical data most closely.

The class exercise invites the students to match the Los Angeles data with any other single factor. In recent years, the factors that most closely matched the data, and in each case did so much better than the model used by the consultants for Los Angeles, were:

• divorces in France,

• egg production in New Zealand, and

• the prison population of the State of Oregon!

Of course, this is all nonsense, which is the point of the exercise. By themselves, good statistical matches of annual data mean very little, if anything, and it is unrealistic to use these procedures to develop and select useful forecasts.

phenomenon. From a practical point of view, we simply cannot have much confidence in the results of analysis that tries to determine the causes of most trends we would like to project.

Recommended Procedure

Recognize Uncertainty

The starting point for good design is the fundamental reality that we cannot predict the future precisely. Our best thinking and our most thoughtful analyses cannot overcome the unpleasant facts that make forecasts unreliable. Interpreting the historical record is ambiguous at best. Although trends may persist over the short term, trend-breakers routinely occur over the longer horizon relevant for product and systems design.

We must recognize the great uncertainties that inevitably surround any forecasts. As the examples illustrate, the realistic range of possibilities is often as wide as a factor of two from a medium estimate for innovative, large-scale projects. The range of possibilities is large even for mature, well-established activities; witness the 10 percent deviation from forecast for air traffic for Boston after only 5 years. We should direct an important part of our forecasting effort toward developing a reasonable understanding of our uncertainties. What is the range of possibilities we may have to deal with over the life of the system or product?

Sometimes we can reasonably assume a symmetric distribution about the forecast of a most likely future, for example, "plus or minus 20 percent." However, the distribution is often significantly skewed. For example, the most likely long-term price of copper is probably in the range of US$1 per pound. Its lowest price will be well above zero, a drop of perhaps US$0.50, but its highest price could be US$4 or more, as has already happened. The distribution of copper prices is thus skewed from the most likely forecast—perhaps only minus 50 percent but possibly plus 300 percent.

A good way to develop an understanding of the size and shape of the range of uncertainties is to look at previous experience in the same or a similar area. The historical record for the product or service often gives a good view of the situation. Thus, a range for the price of copper is easily available from records of past prices. Figures 2.5 and 2.6, respectively, show prices over recent 5 years, for example, and back a century.[5] This kind of information gives a realistic perspective on the range of possibilities. For completely new situations, for which there is no historical record, an effective approach can be to look at situations that appear

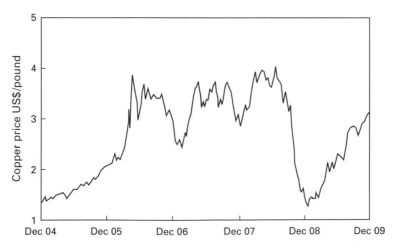

Figure 2.5
Range of copper prices 2005–2010.
Source: www.kitco.com, 2010.

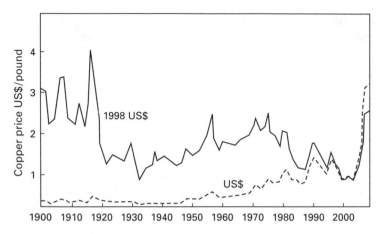

Figure 2.6
Range of copper prices 1900–2010. *Source*: U.S. Geological Survey, 2010.

comparable. We might ask, for example, about the track record of predictions for new electronic technologies when thinking of a new system in that field. We might look at previous experience predicting the size of oil reservoirs when faced with a new site we have just begun to explore. In all cases, we should ask: What is the best information we can have about the size and range of uncertainties?

Where possible, we need to characterize the distribution of the uncertainties. We will be able to weigh the relative value of different designs more accurately to the extent that we can estimate the probability of different outcomes. In some cases, we might be able to do this easily. For example, we can derive the probability of floods, storms, and other natural hazards unaffected by man with some accuracy from historical records. More generally, we must use sophisticated statistical analyses to estimate the nature of the uncertainties (see appendix D on Forecasting).

Shift Forecasting Emphasis
Developing a realistic view of forecasts requires a double shift in the forecasting process. Most obviously, it is desirable to put effort into the process of recognizing uncertainties, defining their ranges, and estimating the probabilities of the outcomes. This implies a shift in emphasis, away from standard efforts to identify the most likely or best forecasts and point estimates and toward spending more time and money thinking about the range of possibilities. Managers and users wanting the most

realistic—and thus most useful—forecasts require forecasters to identify and characterize uncertainties.

The shift in emphasis involves a shift in mindset and expectations. To the extent that forecasters realistically accept the evidence that trend-breakers and other factors disrupt even the most sophisticated estimates of the future, they must also recognize that it is not useful to try to develop very detailed point forecasts. This also means that they can generally forego the very detailed kinds of analyses on which they currently spend so much effort. It is better not only to put more effort into understanding the nature of the uncertainties, but also to put less into trying to get the best estimates of most likely forecasts.

Clients and users of forecasts also need to adjust their expectations. To the extent that they ask forecasters to prepare precise, point forecasts of far-away situations, they are inviting them to participate in an unrealistic charade. The reality is that things happen and situations change, and we should recognize that we can only expect to anticipate ranges of future events.

Track Developments

All the above implies that we need to track developments and readjust our estimates of the future accordingly. This is the logical consequence, once we recognize that forecasts made at any time are unlikely to be reliable due to trend-breakers and other changes in the environment. To develop designs that will be able to meet current and future needs most effectively, system managers need to be on top of these needs. They cannot rely on long-ago forecasts as the proper basis for design, even if those forecasts were approved through some administrative process. Official approval does not define reality.

To track developments most effectively, the forecasting process can identify leading indicators of change. These factors are most likely to signal changes in trends. For example, the developers of the Iridium satellite telephone system might have identified "sales of cell phones" as a leading indicator of the success of their prospective competition. If they had done so early enough, they might have redesigned their system and avoided much of its financial failure. System managers can track these factors over time to obtain early warning of trend-breakers and other developments that might affect the design. They should reallocate their forecasting efforts not just from point forecasts to understanding distributions but also from a one-time effort to a continuous process.

Take Away

System owners, designers, and managers need to change their approach to forecasting. They should avoid asking for the "right forecast" or a narrow range of possible futures. If they request and pay for such a result, they will get it, sure enough: However, it may be of little value. Forecasts are "always wrong"; the future that actually occurs differs routinely from that predicted.

Users of forecasts need to recognize the inevitable uncertainties. They should thus ask for a realistic understanding of the range of possible scenarios. Specifically, they can tailor their requests for forecasts to:

• De-emphasize the development of precise best estimates—such point estimates are demonstrably unreliable, generally expensive, and a waste of money;

• Develop a realistic assessment of both the range of uncertainties around the forecasts and the possible asymmetry of any distribution;

• Estimate, as far as possible, the probabilities associated with possible outcomes; and

• Track developments over time to keep abreast of the situation and adjust designs and plans accordingly.

3 Flexibility Can Increase Value

Invincibility lies in the defense; the possibility of victory in the attack.
—Attributed to Sun Tzu

Robust design is a passive way to deal with uncertainty.
Flexible design is the active way to deal with uncertainty.

—Anonymous

This chapter shows how flexibility in the design of projects can increase expected value, often dramatically. Flexibility in design routinely improves performance by 25 percent or more. It achieves such results by configuring existing technology to perform better over the range of possible future circumstances. Flexible design enables the system to avoid future downside risks and take advantage of new opportunities. By cutting losses and increasing gains over the range of possible futures, flexible design can improve overall average returns, that is, the expected value.

Improvements in expected value can be both financial and other. For businesses, profits and net present value are likely to be the most important metrics. For government and other public services, expected value can be expressed in different terms. A hospital might focus on heath outcomes, for example. Schools might focus on the number of qualified graduates. The military and emergency services might think about response times. Although we focus here on money, because it is broadly applicable, the measure of expected value can take many forms.

Moreover, flexible design often also reduces initial capital expenditures (Capex). It can thus obtain greater expected value at less cost, leading to substantial increases on the return on investment. Indeed, flexible designs lead to smaller and inherently less expensive initial systems because the flexibility to expand as needed means that it is not necessary to build everything that might be needed at the start. Thus, although a flexible design has extra features that enable it to adapt to different situations, the extra cost of these features is frequently much lower than the savings achieved through a smaller initial design.

Our demonstration of the advantages of flexible design sets the stage for the second part of this book, which provides a detailed presentation of the methods for identifying, choosing, and implementing flexibility to maximize value to the system. We take an example, inspired by an actual development in southeast England. The case is transparently simple, enabling us to develop an intuitive understanding of the results, but nevertheless realistic. It represents in the small the same issues and opportunities as large, complex developments.

We begin by showing that standard evaluations, using a cash flow analysis based on a most likely forecast, can give misleading and incorrect results because of the flaw of averages. As we showed in chapter 2, any analysis that does not account for the possible range of distributions of scenarios can be widely wrong—not only in its estimation of value but also, perhaps more important, in its identification of the best project.

Our example demonstrates that the right kind of flexibility in design gives a project leader three kinds of advantages. It can:

• Greatly increase the expected value of the project or products;

• Enable the system manager to control the risks, reducing downside exposure while increasing upside opportunities, thus making it possible for developers to shape the risk profile. This not only gives them greater confidence in the investment but may also reduce their risk premium and further increase value; and

• Often significantly reduce first costs of a project—a counterintuitive result due to the fact that the flexibility to expand means that many capital costs can easily be deferred until prospective needs can be confirmed.

Example: Parking Garage[1]

Our example concerns a multilevel parking garage an entrepreneur plans to build next to a new commercial center in a region where the population is growing. The developer can lease the land for 15 years, after which the land and the garage revert to the landowner. The cost would be $3,330,000 per year.

The site can accommodate 250 cars per level. Due to variation in traffic, we assume that the average utilization of capacity over the year does not exceed 80 percent of installed capacity (i.e., the effective capacity per level is 200 cars). The construction consultant estimates that given the local conditions it will cost $17,000 per space for pre-cast construction, with a 10 percent increase for every level above two, that is, beyond the ground level and the first level up. This premium covers the extra strength in the columns and foundations to carry the load of additional levels and cars.

The average annual revenue per space used will be $10,000 and the average annual operating costs (staff, cleaning, etc.) about $3,000 for each space. The entrepreneur uses a 10 percent discount rate.

An expert consultant has projected demand for parking spaces over the next 15 years. Based on local conditions, the forecast depends on three parameters:

• *Initial demand* for parking during the first year of operation (750 spaces);

• *Demand growth during the first 10 years* of operation (an additional 750 spaces, leading to a demand of 1,500 spaces in year 10); and

• *Demand growth after year 10* (an additional 250 spaces, leading to a saturation demand of 1,750 spaces).

These demand parameters would typically be estimated using demographic projections and past data from similar projects and then used to define a ramp-up curve to the final demand for 1,750 spaces. Our example assumes this curve is of the form

$d(t) = 1,750 - a * \exp(-bt)$

with a and b chosen to match the year 1 and year 10 demands. This gives the demand curve over time shown in figure 3.1 and table 3.1.

For the purpose of the example, we assume that the only design parameter of interest is the number of levels. So the issues addressed are: What is the value of the project? How many levels is it desirable to build?

We use the example in three steps to demonstrate the value of flexibility in design.

• *Base case*   We look at the standard evaluation process, which evaluates a project based on the most likely forecast.

• *Recognizing effects of uncertainty*   We compare the base case with what happens when we recognize that the world is uncertain. The example shows that we not only value the project quite differently from the base

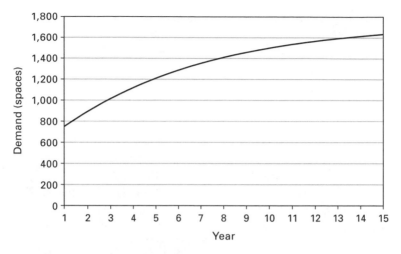

Figure 3.1
Projected growth in demand for parking.

Table 3.1
Projected growth in demand for parking

Year	1	2	3	4	5	6	7	8
Demand (spaces)	750	893	1,015	1,120	1,210	1,287	1,353	1,410

Year	9	10	11	12	13	14	15
Demand (spaces)	1,458	1,500	1,536	1,566	1,593	1,615	1,634

case but also that the best design turns out to be different. In short, we show that the standard process is fundamentally deficient.

• *Using flexible design to manage uncertainty*   We show how the results of the uncertainty analysis help us understand how to improve design and value significantly. In particular, we show how flexibility enables systems managers to adapt the project to the future as it evolves, avoiding downside risks and taking advantage of upside opportunities.

Base Case: Standard Analysis Using Assumed Projection

The standard analysis uses a straightforward discounted cash flow analysis. We set up a spreadsheet of the annual costs and revenues for a specific configuration of the project, sum to find the net cash flows in each period, and discount these sums using a discount rate appropriate to the enterprise. Table 3.2 shows this layout for a particular design. To find the best design—that is, the one that maximizes value—we use this analysis for all the possible levels of the garage. We find out whether the project is worthwhile by seeing whether the best design leads to a positive net present value (NPV) (appendix B gives details on the procedure).

Applying the standard discounted cash flow analysis to the case, we obtain the results in table 3.3. As it shows, the project appears to be worthwhile, delivering positive NPV beyond the required 10 percent cost of capital. Further, the six-level design maximizes the NPV. In this case, as often occurs, the design has a "sweet spot" that maximizes value. The graph of NPV versus number of levels illustrates this phenomenon (figure 3.2). If the garage is too small, it does not generate enough

Table 3.2
Spreadsheet for garage with six levels based on projected demand (costs, revenues, and net present value in $ millions)

Year	0	1	2	...	15
Demand (spaces)	0	750	893	...	1,634
Capacity (spaces)	0	1,200	1,200	...	1,200
Revenue	0.0	7.5	8.9	...	12.0
Operating costs	0.0	3.6	3.6	...	3.6
Land leasing and fixed costs	3.3	3.3	3.3	...	3.3
Cash flow, actual	-3.3	0.6	2.0	...	5.1
Cash flow, discounted	-3.3	0.5	1.7	...	1.2
Cash flow, present value	26.7				
Capacity cost < two levels	6.8				
Capacity cost > two levels	17.4				
NPV	2.5				

Table 3.3
NPV in $ millions of different designs of the parking garage based on projected demand

Number of levels	NPV
4	-1.2
5	2.2
6	**2.5**
7	-0.7

Best performance in each category is highlighted in bold.

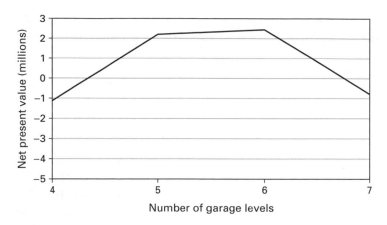

Figure 3.2
The NPV associated with different designs, assuming deterministic demand.

revenue to cover the fixed costs. If it is too large, it may not fill up sufficiently to cover the costs of the large size. The right balance between
being too small and too large lies somewhere in the middle.

This analysis is standard and straightforward. The answers to the basic
questions of evaluation and design appear to be that:

• The best design is to build six levels, and

• The garage is a worthwhile opportunity and delivers an NPV of $2.5
million.

Unfortunately, this solution is wrong on both counts. In the following
sections, we explain why these answers are incorrect.

Recognizing Effects of Uncertainty

Recognizing the Uncertainty

The actual evolution of demand for parking will be highly uncertain over
the 15-year life of the project. Both the initial and additional demands
over the coming decade and beyond could be much lower than expected—
or they could be much higher. Alternatively, there could be low initial
demand and high subsequent growth, or vice versa. The value of the
project under any of these circumstances would not be the same as that
based on the single projected estimate. In general, as indicated in our
discussion of the flaw of averages, we can anticipate that the actual
expected value of the project will be quite different than that calculated

in the naïve case that assumes expected demand as the only possible outcome.

Furthermore, as the example shows, the design that maximizes value can also change when we look at the realistic situation that acknowledges uncertainty. This may or may not happen; however, it is a distinct possibility. As the example indicates, we cannot assume that the ranking of alternative designs will be the same when we consider uncertainty.

To explore the effects of uncertainty in our simple example, we assume that the three parameters defining demand can be off by up to 50 percent either side:

• Initial demand can be anywhere between 375 and 1,125 spaces;

• Demand growth to year 10 can be anywhere between 375 and 1,125 spaces;

• Demand growth after year 10 can be anywhere between 125 and 375 spaces.

As chapter 2 documents, real outcomes can easily diverge from prior forecasts by similar amounts.

We assume for simplicity that none of the parameters in these ranges is more likely than any other. In mathematical terms, we assume a uniform probability distribution of these parameters. The variability in demand thus averages out to the original demand projection. This is important because it ensures a fair comparison between the base case that ignores uncertainty and the more realistic analysis that recognizes it. If the variable demand scenarios had a different average than the original base case, it would not be surprising if the two valuations differed due to a shift in the base case for demand. However, this is not the case in our setup. Any change in the overall valuation of the project will be entirely due to the variability around the projection.

These assumptions lead us to a range of different possible growth curves. Figure 3.3 shows ten of these possibilities together with the original demand projection. Notice the variation in the curves for initial demand and demand growth. In our simplistic model, demand grows over time in all scenarios. This may not be a sensible assumption in reality. More realistic models would account for the possibility that demand might dip. The assumed variability is sufficient to demonstrate the basic point: It is important to recognize uncertainty—if we do not, we get misleading valuations and rankings of projects.

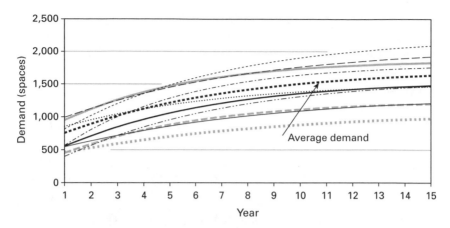

Figure 3.3
Some possible curves of demand growth, compared to original single projection.

Valuation of Project Considering Uncertainty

The valuation of a project considering uncertainty requires a simple piece of technology: Monte Carlo simulation. This is a computer procedure allowing fair consideration of the effects of the possible uncertainties. In practice, the simulation

• *Samples* a large number of scenarios—in our example, thousands of demand curves from the demand distribution;

• *Calculates* and stores the corresponding realized system performance for each scenario, such as the NPV of the project; and

• *Presents* these stored performance data in a range of convenient forms.

Monte Carlo procedures are mathematically solid and reliable. Do not be put off by the association among Monte Carlo, gambling, and irresponsibility. The technology is an indispensable tool for the practical analysis of the effects of uncertainty on design options.

Monte Carlo simulation transforms distributions of uncertain inputs into a distribution of uncertain performance. It does this by repeatedly sampling the uncertain inputs and recording the corresponding performance (see appendix E for details). Importantly, Monte Carlo simulation is efficient. A laptop computer can examine a financial spreadsheet for thousands of possible outcomes in just a few seconds.[2] Monte Carlo simulation thus provides a means of calculating the consequences of uncertain futures.

The immediate result of a Monte Carlo simulation is the evaluation of the expected NPV (ENPV). It is simply the average of all the NPV realizations associated with each of the thousands of simulated possibilities and represents the average value of the project. It is the factor that directly compares with the NPV estimated using a single projection, a forecast of the average conditions. In line with the flaw of averages, and as table 3.4 shows, we typically see that the ENPV differs substantially from the naïve NPV calculated around a single forecast.

Notice in figure 3.4 that the ENPV values for the parking garage are not only different but also are all lower than the naïve estimates based on the single projected demand. Why is this? The explanation in this case

Table 3.4
Comparison of actual ENPV in $ millions and value estimated using single projected demand

	NPV	ENPV
	Based on demand	
Number of levels	Projected	Uncertain
4	–1.2	–2.5
5	2.2	**0.3**
6	**2.5**	–0.1
7	–0.7	–3.8

Best performance in each category is highlighted in bold.

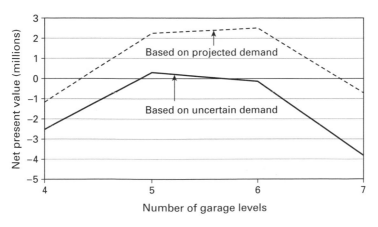

Figure 3.4
Comparison of the actual ENPV associated with uncertain demand and the NPV calculated for a deterministic demand for different designs.

is that higher demands for parking spaces do not increase value sufficiently to compensate for the losses in revenues corresponding to equivalent lower demands. When the parking garage is full, it cannot
accommodate more cars, and it cannot profit from greater demand. This
capacity limitation thus systematically affects the value of the project.

Monte Carlo simulation helps us understand what is going on. The
process generates the information needed to show us in detail what
happens to the system. In addition to providing us with summary statistics, such as the ENPV, it generates graphs of the uncertain system performance. These illustrate what is happening. Figure 3.5, for example,
shows the histogram for the distribution of the NPV for a design of five
levels, obtained from 10,000 sampled demand scenarios. It documents
the fact that although the project generates a distribution of results with
a downside tail of significant losses if demand is low, it does not deliver
a counter-balancing upside tail of higher gains when demand is high. As
figure 3.5 shows, the wide range of possible high demands all lead to the
maximum value the garage can deliver at full utilization. The distribution
of high demands compacts into a narrow range of highest values, thus
giving more than 30 percent probability of being in the highest range, as
the far right of the histogram shows.[3]

It is important to recognize that this skewed distribution of performance can result from balanced distributions of inputs. This is exactly
what happens in this example. The uncertainties equally distributed
around the single forecast have led to the skewed results shown in figure

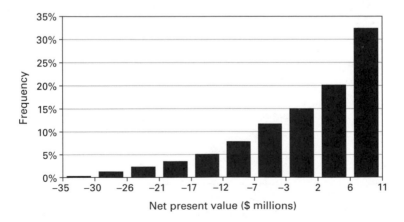

Figure 3.5
Histogram showing the distribution of possible values of the fixed, five-story parking
garage, considering the possible uncertainty in demand.

3.5. The working of the system transforms the data that come in, often into something quite different in character. In this particular case, the capacity constraint of the garage distorts the results downward, as figure 3.6 illustrates. In other cases, the distortion can be upward. In any event, a distortion generally occurs. This is the mechanism that leads to the flaw of averages.

Notice further that the design that maximizes value under realistic conditions is not the same as the one that appears best using a simple forecast. Indeed, the ENPV of the six-story garage, averaged over the 10,000-demand scenario, is negative. The five-story garage has an average NPV of $0.3 million and outperforms the six-story design. The characteristics of the system that change the distribution of the uncertainties can also change the relative value of the design alternatives. As table 3.4 highlights, when we consider uncertainty, the five-level design now appears better—more profitable—than the six-level design that appeared best when looking only at a single forecast!

The standard evaluation procedure thus not only leads to wrong valuations but can also promote wrong designs. How can we identify better designs that deliver best performance under realistic conditions? The answer is: by identifying and exploiting opportunities for value-adding flexible design.

Using Flexible Design to Manage Uncertainty

Understanding the System
A good way to understand the opportunities for improving a design is by looking at the target curve, known technically as the cumulative distribution function. It represents the cumulative chance of obtaining a result below any specific target value, going from the possibility of a result below the lowest value (which has no chance) to a result at or below the highest value (which has a 100 percent chance). Recall that the Monte Carlo simulation had stored thousands of performance levels, corresponding to the thousands of simulated demand scenarios. The target graph simply depicts the percentage of these performance outcomes below the specified target level, hence its alternative denotation as percentile curve.

Figure 3.7 shows the target curve for the six-level parking garage. It indicates, for example, that there is a 10 percent chance of losing about $15 million or more, that is, of missing a –$15 million NPV target. Analysts sometimes refer to this as the 10 percent "value at risk."[4] Reading

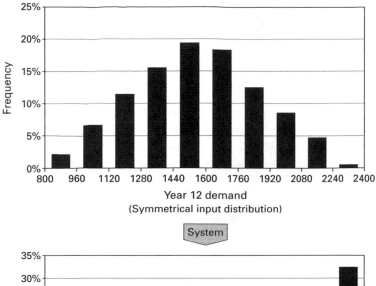

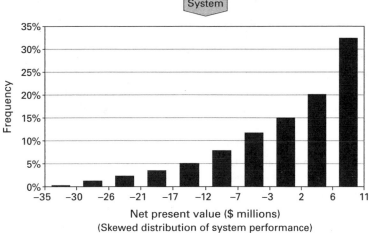

Figure 3.6
Systems distort the shape of uncertainty.

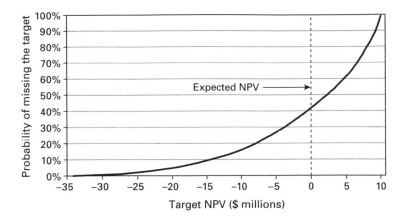

Figure 3.7
Target or percentile curve for the six-level parking garage, indicating the probability of missing any target on the horizontal axis.

up the curve, it also shows that this design has about a 40 percent chance of not breaking even and a 60 percent chance that the realized NPV will be below about $5 million, which is the same as a complementary 40 percent chance of realizing an NPV greater than $5 million.

Instead of a target curve, or chance of missing a target, we may equally display the reverse percentile graph or chance of achieving a certain level of performance. The information in the reverse percentile graph for the parking garage, figure 3.8, is the same as in the target curve in figure 3.7. It shows a 100 percent chance of getting more than the minimum, a 40 percent chance of realizing an NPV of $4 million or more, a 60 percent chance of breaking even, and no chance of exceeding the maximum. Although the percentile and the reverse percentile curves provide identical information, their communication effect can be quite different: People react differently to objectively equivalent probabilistic information. What makes you feel better—an announcement by your doctor that there is a 10 percent chance that you will die or her reassurance of a 90 percent chance of survival? Some people find the reverse percentile graph more intuitive. In some industries, it is the standard way to present data.

We can develop our understanding of the opportunities for improving the expected value of a project by looking at how different designs affect the target curve. Consider figure 3.9, which shows the target curves for designs with different numbers of levels. In the lower left-hand side, the minimum performance increases substantially as we use smaller designs.

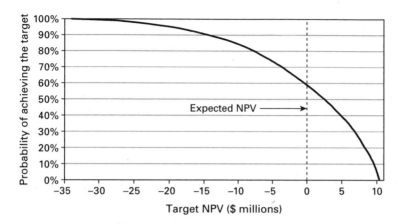

Figure 3.8
Reverse percentile curve, showing the probability that the system will meet or exceed the performance indicated on the horizontal axis.

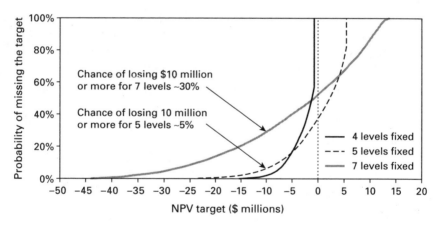

Figure 3.9
Target curves for designs with different levels.

Specifically, the probability of losing $10 million or more drops from 30 percent for a seven-level garage to about 5 percent for a five-level structure. The graph reveals this as a definite improvement.

More generally, *shifting the target curves to the right improves performance*. Moving the curve to the right gives us a higher chance of achieving higher targets. This is a crucial paradigm for the design of systems under uncertainty. We need designs that achieve this improved performance.

Now that we have found a way to articulate improved performance under uncertainty, we can try to understand why we get this result. Developing our understanding of how the system responds to uncertainties will help us develop solutions that maximize performance. In this case, the reason that smaller designs are less likely to lose money is clear: If you invest less at the start, you reduce the maximum amount you can lose. Of course, this does not mean we should always invest less; spending more—for example, in enabling flexibility—is often the way to increase overall expected value.

Phasing Investments

We can phase investments to reduce initial cost and therefore the maximum exposure to loss. The phasing can either be fixed, in that it follows a predetermined schedule, or flexible, in that the later stages are optional.

Phased design can be a good idea even without the recognition of uncertainty. Deferring the cost of building capacity until needed generally saves on financing costs. At a reasonable opportunity cost of capital, 10 percent for example, deferring capital expense by 4 years reduces the present value of each deferred dollar to 70 cents. However, the economies of scale that designers might achieve by building capacity all at once counterbalance the potential savings of phasing to some degree.[5] Moreover, financing and other costs may vary over time. The value of phasing needs careful consideration.

Based on the projected demand growth for the parking garage, it would be reasonable to plan a fixed-phased expansion. The idea would be to add floors one by one, tracking the projected demand as follows:

- Begin with four levels,
- Expand to five levels in year 3,
- Expand to six levels in year 4, and
- Expand to seven levels in year 7.

Note that phasing requires the original design to incorporate the ability to expand. In this case, planned phasing to seven levels means that we would have to build the first four-level phase with extra-large columns and foundations sufficient to carry the eventual loads of a higher structure. We assume this cost adds 30 percent to the cost of the first two levels.

From the perspective of a standard NPV analysis based on projected demand, this fixed phasing more than doubles the NPV of the project, from $2.5 million for a fixed six-level design to $5.4 million for the phased design (see table 3.5). From the naïve perspective, this fixed phasing seems highly attractive. But how does it compare when we value it recognizing uncertainty?

The ENPV for the fixed-phased design when we recognize uncertainty is greatly different from the value based on the single projection. In fact, when we test the design with 10,000 demand scenarios, the realized ENPV averages at –$0.4 million (see table 3.5). This realistic assessment shows that this fixed-phased design is actually worse than the fixed designs of five or six levels—contrary to what the standard analysis indicated.

To understand what is happening, we can look at the target curve comparing the fixed-phased design with the better five-level fixed alternative (figure 3.10). The fixed-phased design curve is far to the left for lower values. This means that it is much more likely to lead to major losses. This is because the fixed-phased design locks into an eventual larger garage—up to seven levels. If demand increases as projected, this plan leads to good results. Hence, the target curve for the phased design

Table 3.5
Evaluation of phased demand, values in $ millions

	NPV	ENPV	
	Based on demand		
Number of design levels	Projected	Uncertain	Break-even probability percent
4	–1.2	–2.5	0
5	2.2	**0.3**	64
6	**2.5**	–0.1	59
7	–0.7	–3.8	48
Phased	**5.4**	**–0.4**	65

Best performance in each category is highlighted in bold.

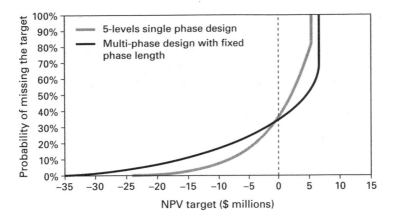

Figure 3.10
Target curve comparing phased design with fixed design with five levels.

curve is further to the right than the corresponding part of the fixed design curve. However, if demand is less than expected, the commitment to build large is obviously wasteful. Locking the system into an expansion plan is not a good idea when the need does not materialize as initially expected.

We can use the target curves to identify the probability that any particular design will break even, that is, repay the present value of the investment costs. This break-even probability is the complement of the probability that the project has an NPV of zero or less. Thus, table 3.5 shows the break-even probability for the phased design to be 65 percent, which is 1 less the intercept of its target curve with NPV = 0 in figure 3.10.

Flexibility Is the Way

The problem with the fixed-phased design is that it locks into an expansion plan that may not make sense. This observation points to the preferred solution. A flexible design permits *but does not require* expansion. A flexible design positions the system so that it can expand when it makes sense to do so but does not commit managers to expansion if the actual situation that occurs does not justify greater capacity.

A flexible design recognizes that we will learn more about system requirements during the lifetime of the project. Once we open the system, we will quickly know the initial demand. As we gain experience, we will also be able to get a better idea of demand growth. Consequently, we may postpone or accelerate the expansion; we may expand less or not at all, or may expand more.

Proper development of a flexible design requires reasonable criteria for expansion. Depending on the situation, managers may want to expand to a certain extent in anticipation of demand to avoid lost sales. Alternatively, they may want to defer expansion until a pattern of higher demand is well established. Judgment about which criteria to apply is a management decision. From an analytic perspective, we can easily program any management decision rule into the analysis.[6] We can therefore anticipate the range of outcomes and the future performance of the system.

Flexible design can indeed be far superior. Consider, for example, the flexible design of the garage with an initial build of four levels that incorporates the ability to expand incrementally up to nine levels. To value this design, we have to agree on a criterion for expansion or a contingency plan. In this example, our decision rule was to add another level whenever demand exceeds capacity in the previous 2 years. As table 3.6 shows, the flexible design is preferable on all counts to the fixed alternatives.

The flexible design delivers much greater ENPV than the best alternative fixed design, $2.1 million compared with $0.3 million. This kind of result is common, as we demonstrate with case examples throughout the rest of this book. By building small, flexible design lowers the risk of bad performance: The less that is invested, the less that can be lost. By incorporating the ability to expand easily, it allows system managers to take advantage of favorable opportunities whenever they appear. Both actions lead to greater expected value. They lower the value at risk and

Table 3.6
Comparison of flexible design with alternatives, values in $ millions

Number of levels	Initial capital outlay, Capex	NPV Based on demand Projected	ENPV Based on demand Uncertain	Break-even probability percent	10 percent value at Risk	to Gain
5	19.2	2.2	**0.3**	64	–7.6	5.4
6	24.2	2.5	–0.1	59	–14.3	9.6
4, flexible expansion	16.7	n/a	**2.1**	66	–7.4	9.2
Which best?	Flexible	n/a	Flexible	Flexible	Flexible	Flexible

Best performance in each category is highlighted in bold.

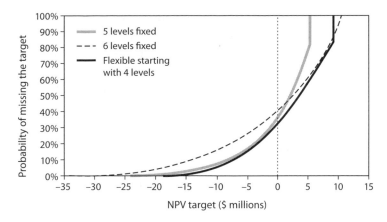

Figure 3.11
Target curve for flexible design compared with the best fixed alternatives.

Table 3.7
Summary of results for parking garage example, values in $ millions

Number of levels	Initial capital outlay, Capex	NPV Based on demand Projected	ENPV Based on demand Uncertain	Break-even probability percent	10 percent value at Risk	to Gain
4	14.7	–1.2	–2.5	0	–6.3	–0.7
5	19.2	2.2	0.3	65	–7.6	5.4
6	24.2	2.5	–0.1	60	–14.3	9.6
7	29.6	–0.7	–3.8	45	–23.5	10.7
Phased, planned	16.7	5.4	–0.4	65	–16.2	6.6
3, flexible	12.6	n/a	0.8	60	–5.9	5.2
4, flexible	16.7	n/a	**2.1**	65	–7.4	9.2
5, flexible	21.2	n/a	1.6	60	–10.8	11.5

Best performance in each category is highlighted in bold.

raise the value to gain. The target curves in figure 3.11 show this effect graphically (see table 3.7 for a summary). The NPV target curve for the four-level flexible design mirrors the performance of the five-level fixed design in low-demand scenarios but moves to the curve for the six-level fixed design in high-demand scenarios. Avoiding losses and increasing possible gains moves the target curve to the right, toward better performance.

Most remarkably, *flexible designs often cost less than inflexible designs.* This result is truly important and runs against the intuition that flexibility always costs more. To appreciate this fact, it is important to create a fair comparison between the flexible or inflexible designs that we might implement. The basis for comparison is our initial capital investments (Capex). The comparison needs to be based on what we might actually decide, not what the initial structure looks like. Thus, we need to compare the following possible decisions:

• The choice of inflexible design means that we have to build it larger than immediately needed so that we can benefit from future growth—the five-story solution; and

• The choice of flexible design allows us to build only for immediate needs, as this design allows us to add levels as needed—the four-story solution.

In short, the fair comparison is between the alternative investments, not between two 4-story and two 5-story designs. The fair comparison is between the four-story flexible and the five-story inflexible designs.

For this case, the initial cost of the flexible design is 10 percent less expensive than the competitive five-level design. This is because flexible designs allow system managers to build small initially, which can create tremendous savings—far greater than the cost of enabling the flexibility. In this case, the extra cost of building stronger columns and footing to enable expansion is less than the savings resulting from building the structure smaller initially. Flexibility can be a "win-win" design!

Similar benefits occur in major projects. Figure 3.12 shows the increased benefits associated with alternative flexible designs for major automobile factories in the United States. The situation here is that manufacturers have to equip their plants several years ahead of production of new models; it takes time to order, obtain, and install the specialized equipment necessary for each type of car. Meanwhile, the level of future sales for the different models and products is highly uncertain. Yet major manufacturers have so far been designing the plants around their estimates of the most likely demand, much as we showed in the garage case. The target curves in figure 3.12 illustrate how flexible designs can greatly increase ENPV.

The parking garage case is a simple example, but its principles are widely applicable. A large-scale project that is close to the case example is the staged development of the Health Care Service Corporation (HCSC) building in Chicago, described in chapter 1.

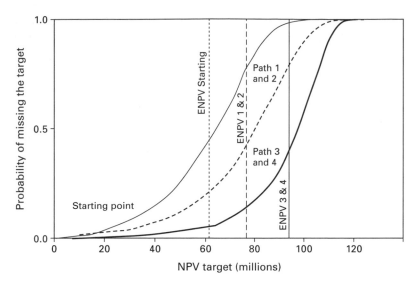

Figure 3.12
Target curves for flexibility in the design and equipment of major automobile factories.
Source: Yang, 2009.

Take Away

Flexibility is an effective way to improve the expected performance of systems in uncertain environments. The gains to be made can be impressive. This is especially so when the flexible design also reduces the initial capital expenditure required for the project.

The types of flexibility that are most useful in improving performance depend on context. Product or system designers face the challenge of using their technological expertise to create appropriate systems that can cope with many unfolding futures.

Valuing projects correctly requires the recognition of uncertainty. Evaluations based on average or most likely forecasts of future needs or requirements will systematically lead to incorrect answers with regard to both the level of value and the ranking of design choices. Proper analyses require an understanding of the effects of uncertainty.

Target curves representing the uncertain performance of a system highlight the risks and opportunities of different designs and provide an effective, systematic tool to gain insight into the behavior of alternative designs. They give meaning to the notion of "improving" the uncertain performance of a system: The goal is to use creative design to move the performance curve, or at least substantial parts of it, to the right, to reduce downside risks and increase opportunities for benefiting from the upside.

Monte Carlo simulation is a widely available practical tool that allows designers and economists to calculate distributions of the future conditions within which their project will develop. It enables us to calculate the realistic expected value of a design and to produce target curves and other information that fully describe the consequences of alternative designs.

II METHODS OF ANALYSIS

... in this world nothing can be said to be certain, except death and taxes.
—Benjamin Franklin (1789)

In the first part of this book, we gave an overview of the value of flexibility in design. First we presented the reason why it is necessary to recognize uncertainty explicitly to avoid erroneous calculations and inferior results. We then show the benefits of suitable flexible designs, which lead to great improvements in expected value compared with designs based on deterministic forecasts. Our aim in part I is to provide an overview that will be useful to all readers, specifically busy leaders at the top of the organizations involved in large-scale technology projects. We hope we have convinced the latter of the need to have their teams recognize uncertainty explicitly and look for good flexible designs.

In part II, we provide more detailed descriptions of the major techniques for identifying good flexible designs. We assume the reader is conversant with basic analytic approaches to complex issues, such as modeling, statistical analysis, Monte Carlo simulation, and discounted cash flow analysis. Appendices on these and related topics complement the chapters in part II. The second part of this book is therefore for current or future implementers of flexible design—senior engineers and managers, junior staff, or students aspiring to leadership in system design.

Our goal is to enable you to ask the right questions, direct your staff and consultants in the right direction, and interpret and communicate the results of their technical analyses to a non-technical audience higher up or outside your organization. We also aim to help you carry out some "back-of-the-spreadsheet" prototype modeling yourself to help you understand on a small, simplified scale what goes on in the large-scale analyses you will be asking your staff to undertake. Conversely, we do not aspire to train you to become a statistician or an expert mathematical modeler. We assume that you will delegate routine technical aspects of the work to specialists in your team or consultants with appropriate expertise. We intend part II to enable you to have a meaningful discussion based on solid analysis, with modeling experts, on the one hand, and senior leaders and stakeholders, on the other hand.

Analytically, part II presents a range of new approaches suitable to analysis and evaluation of systems under uncertainty. Our objective in this respect is to enhance traditional modeling techniques, not to replace them. In each chapter, we build on the established methodology—for example, a discounted cash flow model—and enhance it in such a way that any new features can be "turned off" to recover the traditional model. Our intention is to make it easy for practitioners to adopt the concepts of flexible design by integrating new concepts easily into established practices.

Our approach is pragmatic, directed toward practitioners who will be designing, managing, and implementing real projects. We aim to improve practice as much and as easily as possible. More academically inclined colleagues may think we should instead be presenting theoretically correct methods that "get it right." However, we believe that our modest aspiration to "get it better" is more likely to improve practice. Indeed, the concept of "getting it right" is difficult to defend once we accept that modeling the performance of socio-technological systems is as much an art as a science.

In detail, part II covers the main elements needed for the development, selection, and implementation of flexible designs. We present a systematic process to follow for each topic. Our aim is to provide a coherent framework for integrating the many relevant analytic approaches involved. It is organized as follows:

• Chapter 4 focuses on estimating the distribution of future possibilities. This chapter deals with the issues associated with recognizing the uncertainties around complex projects. It extends the usual methods of forecasting, which seek to develop the most likely outcome, to procedures for developing a range of possibilities that might occur. Getting a good vision of the range of what may actually happen is essential to developing designs that will deal effectively with what the future brings.

• Chapter 5 develops ways to identify candidate flexibilities. It provides a process for identifying the kinds of flexibility that may be most useful and can add the most value to a project, something that is generally not obvious in a complex system. This section shows how we can use effectively simplified models of our systems in combination with Monte Carlo simulation to identify productive forms of flexibility.

• Chapter 6 presents effective ways to evaluate and select preferred designs. It builds on standard discounted cash flow analyses to provide graphical and tabular presentations of the multidimensional trade-offs designers and managers inevitably have to make among increased value, higher risk, and other criteria of good design.

• Chapter 7 focuses on ways to increase the possibility of effective implementation of flexibilities in design. It provides a notional checklist of steps that designers and managers can take to increase the likelihood that they will be able to use the flexible capabilities designed into a system when they want to do so.

• Chapter 8 is an epilogue. It closes part II with an overall assessment of what we can already achieve in flexibility in design and of the further practical work we need to do to improve our capabilities and procedures. Although basic principles and numerous case examples indicate that flexible design can greatly improve expected value, the systematic approach proposed in this book is new, and we can certainly improve it.

4   Phase 1: Estimating the Distribution of Future Possibilities

I mistrust isolated trends.... In a period of rapid change, strategic planning based on straight-line trend extrapolation is inherently treacherous.... What is needed for planning is not a set of isolated trends, but multidimensional models that interrelate forces—technological, social, political, even cultural, along with the economics.
—Alvin Toffler (1985)

The only constant is change, continuing change, inevitable change; that is the dominant factor in society today. No sensible decision can be made any longer without taking into account not only the world as it is, but the world as it will be.
—Isaac Asimov, building on Heraclitus (5th century BCE)

In this chapter, we focus on the challenge of estimating what may happen, which is indeed a challenge. We have to recognize that unpredicted events shape both our lives and that of our projects. Things happen. New technologies change our world, as penicillin radically increased life expectancy and affected population growth. Political events upset expectations and demands: Look at how the creation of the European Union has reshaped markets and standards over the continent. Economic booms and busts create and reduce demand for services and products drastically compared with previous trends. Nonetheless, we still need estimates we can work with to design projects and systems. We need somehow to move ahead while being modest about how accurate we can be. We do not have the omniscience to predict precisely what may happen. We need to be satisfied with good estimates of the distribution of possibilities, of the range of what may happen and the relative likelihoods of various scenarios.

We need two kinds of estimate of future distributions, as figure 4.1 suggests. One concerns the external circumstances that drive system performance, for example, the economic environment and demand for

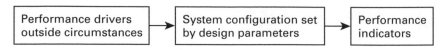

Figure 4.1
Elements in a chain of uncertainty in a system.

products, the price of oil and other commodities and materials, prevailing regulations, and so on. These circumstances generally do not depend on the operation of the system we are designing. The other kind of estimate relates to the performance metrics of the system itself, such as the level and cost of production. Estimates about system performance require an understanding of the system's design and operation; they require an adequate model and a meaningful description of its operation and dynamics. Designers and managers have to adopt different approaches for each case.

Experts other than system designers usually generate estimates of outside circumstances. For example, economists will generate forecasts of economic activity, market analysts will estimate overall demand for products, and so on. Similarly, geologists normally provide estimates of possible quantities of ore in a mine or of petroleum in a field. System designers need to understand the assumptions behind these forecasts, and they need to ask the right questions so they can obtain the most appropriate estimates for effective system design. These estimates should recognize a range of possible future paths. In practice, we can place these probabilistic estimates in spreadsheets that provide inputs to a sensitivity, or Monte Carlo, analysis that explores the possible effect of various scenarios (see appendix D).

Designers and managers have a more difficult task when it comes to estimating the distribution of the performance of their own systems. Before they can carry out the forecasting exercise, they need to develop a suitable model of their system. To obtain good performance estimates, they must link the interaction of external circumstances with their design to generate key performance measures for the system. Good models of systems combine both technical and social considerations (such as economics). Thus, their development can be a special challenge for the professional experts in the system. Design engineers almost never have studied economics seriously. Doctors who are experts in their specialty and understand the functioning of their unit very well are rarely trained or otherwise prepared to understand how their work affects the hospital's overall performance, as measured by costs, average delays in service, or patient readmission rates. System designers and managers need to pay special attention to the development of useful models of their operations.

Note that the emphasis on estimating distributions considerably extends current practice. The concern with the range of possible outcomes requires a realistic appraisal of the range of possible design inputs and requirements. However, designers are traditionally educated to work with fixed specifications and generally prefer to do so. It makes their work easier. With fixed specifications, they do not have to worry about many different combinations of events. Indeed, accepted practice for system design starts with the definition of fixed requirements.[1] This approach calls for specific, precise forecasts of future contexts and needs, known as "point forecasts." Unfortunately, such point forecasts are generally "wrong," in that the actual results are different from best estimates (see box 4.1).[2] The realistic approach is different; we cannot know the future precisely; we can only estimate it as a range of possibilities.

Moreover, this traditional approach focusing on most likely futures leads to faulty estimates of performance and value, even if the estimates of future conditions are correct on average. This is due to the flaw of averages described in chapter 1 and appendix A. A crucial message of this book is that the traditional approach ignores the reality that the system will operate in a wide range of circumstances and makes no provision for adapting the system to these conditions. The traditional approach does not incorporate flexibility and fails to develop the value it provides. To access the increase in expected value we can obtain through flexibility,

Box 4.1
Examples of point forecasts that served as the basis for major designs

• In 1988, forecasters predicted that there would be 1 million customers for satellite telephone services in 1998. This was a driving criterion for the design of the Iridium satellite communication system launched a decade later. In 1998, Iridium had only about 50,000 customers (see box 1.1).

• In 1982, airport planners for the new Lisbon international airport projected that it would serve 23 million passengers in 2010. By that date, the traffic was only slightly over half that prediction, and the government seemed unlikely to support a new airport built to those plans (NAER, 1982).

we must focus on estimating distributions of outcomes, not single number projections.

Estimating the distribution of future possibilities is a five-step process:

1. *Identify the important factors*   The first step focuses on the factors that are most crucial for the future performance of the system we are to design. Because the number of different variables that can affect the performance of any large-size system can be huge, it is crucial to prioritize effectively to reduce their number to a handful that are mission-critical for future performance. This prioritization requires expertise and insight into the wider operation of the system from engineering, economic, and management perspectives.

2. *Analyze historical trends*   The idea is to establish the historical trends of the key performance drivers and the historical variability around this trend. What do we know about the situation today? How did we get where we are? In the early phases of planning a large system, we hear many speculative arguments that attempt to explain how the current situation emerged. These often stem from anecdotal evidence, colored by personal experience and responsibility. It is important to use as much hard data as possible to challenge unfounded assumptions that might otherwise distort the analysis.

3. *Identify trend-breakers*   What are the potential trend-breakers? How likely are they to arise? How long will historical trends continue? Responses to these questions are more speculative, based on judgment rather than evidence, and therefore prone to biases. We can minimize these biases through careful analysis of historical data—by impelling people to base their arguments on facts and by establishing and managing a constructive scenario-planning process that will challenge assumptions. It is important to involve major stakeholders in scenario planning, both those accountable for decisions and relevant domain experts, for example, doctors in a hospital. The idea is to develop a realistic set of future scenarios so that the design can anticipate these eventualities.

4. *Establish forecast (in)accuracy*   We also need to be realistic about our collective ability to predict accurately in our particular field. A priori, our past performance provides a good indication of how well we can hope to execute our next project. If an automobile company's previous sales forecasts for new models have differed from reality by X percent on average, this represents a reasonable estimate of its likely error in the forecast sales for the next model. We should document our record of accomplishment by comparing previous forecasts with what actually happened. As

chapter 2 indicates, our forecasts are "always wrong," in that there are normally significant discrepancies between what we or our predecessors predict and what actually occurs. It is surprising how few companies store, let alone systematically analyze, their past forecast errors.

5. *Build a dynamic model*   Finally, we put these four steps together and build a model, or several models, that allow us to generate many different possible evolutions of the system environment. These dynamic models replace the traditional single-number projections of input variables. Engineers and system architects should optimize their designs against the fluid and unpredictable futures that these models produce.

Throughout the rest of this chapter, we develop these steps sequentially and illustrate them with a range of examples. To demonstrate the entire process, we end with a case study of an important practical example involving the development of major hospital facilities.

Step 1: Identify Important Factors

Identifying the key performance drivers of a system is not as easy as it seems. The obvious answers are often insufficient. For example, developers often present the need for a project in the following terms: demand for some product is growing, soon it will exceed the available capacity, and therefore we need additional capacity. This is the classic argument.[3] Promoters routinely use this reasoning to justify all kinds of public and private projects, such as airports, factories, highways, power plants, steel mills, and so on. The obvious performance driver of the system thus seems to be aggregate demand—for air travel, factory products, and so on. This understanding is not wrong, but it is insufficient.

The obvious identification of aggregate demand for a service as a performance driver is inadequate because it does not recognize that demand inevitably comes in many forms, which are unequal and have different implications for system performance. For example, the demand for power in off-peak and peak hours has very different implications for both the design and performance of an electric power plant. A nuclear reactor is very efficient at meeting continuous demands that continue day and night (so-called base loads) but cannot ramp up its production quickly to meet peak loads. Conversely, turbines driven by diesel engines can deliver power quickly for peak loads but are uneconomical for providing base load power. In short, demand is complex. We need to define such drivers in detail.

In determining useful performance drivers, we should be concerned with three elements:

• *Type*   What are the elements that will affect system performance?

• *Level of aggregation*   What is the best way to specify each type?

• *Variability over time*   How does the driver fluctuate? How does this uncertainty enhance the value of flexibility?

We also need to decide:

• *Criteria of usefulness*   Which of these variables do we really need to include in the analysis?

We now discuss how to deal with each of these issues in turn.

Type of Driver
We need to identify the factors that influence system design and performance. These may be economic, technical, regulatory, and others. In general, they may be much broader than system designers initially imagine before they think hard about their project. There is no convenient checklist. We need to think deeply about our system and identify the factors that will influence its performance. This process requires the collaboration of a range of experts. The objective of this phase is to develop a primary list of important drivers for consideration.

Consider the development of a power plant, for example. As we suggested above, the various forms of demand for power (peak and off-peak) strongly influence the desirable design and value of a plant. Technical uncertainties will come into play, too—the resistance of the soil to earthquakes, for instance. Possible future regulations about the need for carbon capture will also influence design and system performance. Of course, identifying the important drivers will depend on the circumstances. From 2010 in Europe, regulations about carbon emissions—and uncertainties about their level—became major drivers of power plant design. This was not the case in Europe a decade earlier or in China at the same time.

Level of Aggregation
It is often useful to think in more detail about drivers, as suggested by our discussion on types of demand. When designing a power system, it is crucial to recognize that the demand for power at different times has significant implications for what society builds. Moreover, the demand

Box 4.2
Identification of variables for airport planning in 2010 in Europe

The traditional drivers for airport development are the number of aircraft operations and passengers separated into international and domestic where applicable.[14] These drivers obviously connect directly to the design of the airside (runways and taxiways) and landside (domestic and international terminal facilities) of an airport.

A little thought indicates that designers should consider these drivers in more detail. On the landside, we should distinguish between passengers flying with traditional and low-cost airlines because these types of airline require different kinds of terminal building. The low-cost airlines typically want simpler, cheaper facilities. On the airside, we should distinguish between airlines operating many connecting services (hub airlines) and those that do not (point-to-point airlines). Connecting airlines want to concentrate arrivals and departures together to reduce connecting times for passengers and thus provide convenient service, whereas point-to-point airlines spread their flights more evenly over the day. In this way, the type of airline drives the amount of runway capacity needed.

Similarly, in some circumstances, airport designers may want to think of drivers at a more aggregated level. In regions where airlines are merging, such as Europe in the early 21st century, designers need to consider the possibility that a major airline might disappear, which would greatly affect the need for airport facilities.[15]

for power comes in different forms; higher voltages are necessary for transmission and some industries, lower voltages for household consumption.

It can also be useful to think of higher levels of aggregation. When considering the development of an oil field, for example, it is clear that both the price of oil and construction costs affect the overall profitability of a project. However, the overall economic situation may affect both: Boom times increase the demand and thus the price of oil; this in turn increases demand for drilling rigs and pipelines and the cost of building platforms. In some circumstances, it might be useful to focus on some universal drivers, such as the state of the overall economy in this case.

Box 4.2 illustrates the identification of performance drivers at various levels of aggregation.

Variability over Time

It is also important to disaggregate drivers, particularly when drivers may vary significantly over time, seasonally, for example. Demands for power

Box 4.3
Vancouver: Flexible design for variable airport traffic

Conventional design for airport terminals creates facilities to meet a reasonable peak for several kinds of drivers.[16] Airport designers may calculate the size of facilities to meet the needs of international and domestic passengers and sum up these spaces to plan the overall building size.

Better practice recognizes that the peaks of international and domestic passengers can occur at different times. Domestic passenger traffic often peaks during the morning and evening rush hours, whereas intercontinental traffic peaks in mid-afternoon or late at night due to the constraints of time zones. Designers can thus save significantly on the overall size of the building if they make a significant fraction of the space flexible to serve domestic passengers when their flow peaks and international, intercontinental passengers during theirs. Creating this kind of "swing" space reduces the overall requirements by a third or more, thus correspondingly increasing design efficiency.[17]

Vancouver airport in Canada is notably efficient in using flexible space for international and domestic passengers. It uses glass partitions and doors that allow aircraft gates or passenger lounges to be secured for either use.

at peak times of day in peak season may drive the determination of the capacity of a power system.

Because variability creates opportunities for flexibility, careful consideration of variability over time is particularly important when several different drivers require similar facilities. If the peak demands for different services occur at different times, facilities for the peak requirements of one service may be available for another at its peak. This flexibility can be a great source of improved, more efficient design, as box 4.3 illustrates.

Criteria of Usefulness

Pragmatism is essential when developing a final choice of variables. We need to keep in mind that our objective is to perform an effective analysis that will help us understand relevant issues and develop improved designs. For this reason, we should focus on a set of a few, simple variables that we expect to have impact.

Most importantly, the uncertainty level of the variables selected as drivers should have significant design consequences. If that is not the case, those variables should not be a priority for consideration at this stage. If we look at the airport example in box 4.2, we see that there is uncertainty in the proportion of travelers who do not speak the local language. This information may be useful when it comes to thinking about interior signage, but it is highly unlikely to influence the overall design of the system. It is not the sort of variable that will be used as a driver for the design of an airport terminal.

The fewer variables the better, otherwise the analysis for the overall design will become too complicated. The fact is that the number of scenarios to be considered increases exponentially with the number of driving variables. That is, if we want to consider three levels of one variable and three of another, then we have to think about 3 x 3 = 9 combinations. This means that we can only properly consider a few variables as drivers of design, even with our fast modern computers. We elaborate on this important feature in chapter 5. There are tools, such as Tornado diagrams (explained in appendix D on Monte Carlo simulation) that help prioritize a long list of uncertain drivers. As with all tools, we need to use them to complement, not substitute for, expert judgment.

Finally, it is crucial to keep the choice of variables simple and intelligible. To be able to make good judgments about a design, we need to be able to develop an understanding of how it works. Complex technical factors can prevent us from forming a clear idea of what causes what and planning a coherent design. Designers have to be able to explain and justify their project to decision makers who will be unfamiliar with the details. They will need to be able to make a convincing argument that others will understand and trust. If the model of the system is too complex, neither designers nor managers will find it acceptable.

Step 2: Analyze Historical Trends

The purpose of analyzing historical trends is to develop our understanding of what has been happening, with a view to estimating future possibilities. A proper analysis should go through several phases, each building on the other. We need to:

- *Understand the data* How good are they? What do they represent?
- *Develop an appreciation for the overall pattern* What is the central line?
- *Assess the uncertainty in the trend* How steady has it been? How has it fluctuated away from the central line?

Once we have completed these phases, we will be in a position to estimate future distributions, through an examination of how trends may break, and then more generally of the several factors that shape the evolution over time.

Understand the Data

The first part of the analysis of historical trends is to assess the nature and quality of the data. Institutions record data for a variety of purposes, under definitions, procedures, and circumstances relevant to their immediate situation. These conditions may not be appropriate for planning and designing a facility.

We should first check the definitions, which often do not mean quite what we think. For example, the U.S. Federal Aviation Administration (FAA) keeps a record of aircraft delays. However, this measure only represents a small fraction of what passengers might consider a delay. By convention, the FAA only starts counting delays once they are more than 15 minutes over the scheduled time. Moreover, knowing that "on-time performance" is one of their metrics, the airlines pad their schedules to account for all the time their aircraft have to wait for takeoff, landing, and so on. So the scheduled flight time from Boston to Washington is now about 30 minutes longer than it was 30 years ago. This increase is due to the time aircraft have to wait or slow down due to crowding and lack of capacity, which constitutes real delay. Yet despite the reality of delay in the system, flights can be "on time" and so have no official delay. Additionally, of course, definitions change and often differ from place to place and between organizations.[4] The bottom line is that it is useful to discuss the meaning of data with relevant domain experts.

We also need to check data quality. Data may contain errors or be systematically biased (see box 4.4). In the United States, doctors appear to have systematically under-recorded deaths from AIDS, just as they did deaths from alcoholism in the early twentieth century because of the social shame associated with these conditions. On the other hand, people whose wages depend on the amount of work they do have been known to inflate their workload records.

Develop Appreciation for Overall Pattern

The standard process for estimating data trends over time is a regression analysis, a process for finding the line or equation that best fits the data that is regularly available as a function in a spreadsheet. Regression

Box 4.4
Interpreting data: Water use in Boston

Designers based the original plans for the $6 billion sewage treatment facilities in Boston harbor on historical data from the Metropolitan Water Resources Agency (MWRA). These data showed steadily increasing consumption of more than 300 gallons (1,100 liters) of water per person per day. Accordingly, the original designs extrapolated future daily use of about 350 gallons per person.

Did the data mean that Bostonians actually washed, drank, and flushed that much? Actually, not. They meant that the MWRA sent that much water from the Quabbin Reservoir into the Boston system. However, about one third of this water seemed to have been lost through leaks in the old distribution system. The sewage plant therefore could be much smaller than initially planned.[18]

Did the trends demonstrate that usage would continue to grow? Again, no. Once the authorities built the facility, they raised the price of water to cover the costs of construction. Households went from paying about $20 to around $500 a year. Higher prices encouraged consumers and industry to deal with leaks and install water-saving devices. The overall consumption of water thus decreased about 20 percent compared with the earlier baseline—to around 70 percent of basis for design.

analysis usually calculates the trend as a straight-line function of time t of the form:

Value of factor at time t = a t + b

The trend may also be exponential to represent a constant rate of growth:

Value of factor at time t = b e ^{at}

If time t refers to years, its count is usually nominal (1, 2, 3…) rather than the actual calendar year. It is generally sensible to use expert judgment to choose a reasonable trend, which can include more complex patterns, such as cycles. Regression is used to determine the parameters, such as "a" and "b" above, by fitting the curve most closely to available data. Appendix E gives examples of regression analyses.

The issue with developing a trend is that the result is sensitive to the period analyzed. If the period considered started in a recession and ended with a boom period, the trend might increase rapidly, going from relatively low to relatively high. If we take a much longer period to cancel out the effect of cycles of activity, we may reach back into times when trends were completely different. In short, the results of a regression analysis depend on our arbitrary choices of length of time and data, as box 2.6 and figure 2.4 in chapter 2 show. The mathematically precise regression analysis inevitably leads to imprecise, assumption-dependent estimates of trends.

Assess the Uncertainty in the Trend

Regression analysis automatically provides an immediate way to estimate the goodness of fit between the trend and the observations: the R-squared metric. By definition, R-squared ranges between 1 (all the observations match the trend exactly) and 0 (there is no evidence that the data grow or decline linearly or exponentially over time).[5] Because this measure is always available, people frequently refer to it. However, it is not reliable. It can easily be increased by adding factors to the statistical analysis, which guarantees higher R-squared, and it is often meaningless because many observations grow exponentially, and so tend to fit very well with each other even though they have no real connection. The R-squared metric is therefore not a good indicator of the usefulness of a trend.

We can generate a more useful measure of variability of the trend line by carrying out regression analyses for data over different periods. The result is a set of different trend lines that provide an immediate sense of the range of possibilities. Figure 4.2 shows the kind of results possible. Such efforts provide a clear indication of the uncertainty in trends.

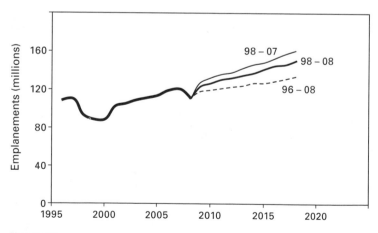

Figure 4.2
Trends estimated by statistical analysis are sensitive to the period chosen for analysis.
Source: U.S. Federal Aviation Administration data for a major airport.

Step 3: Identify Trend-Breakers

The analyses in the previous steps are mechanical. Step 1 validates the data, and step 2 summarizes them in trends using standard statistical analysis. Analysts mostly organize data; they do not have to think about them much. However, step 3 requires careful thinking beyond the mathematical analysis.

It is important to identify "trend-breakers." As we indicated in chapter 2, trend-breakers are events that disrupt the smooth continuation of what has been happening in the recent past. They come in various forms—economic crises, political reconfigurations, new technologies, new discoveries, and new market conditions—and happen routinely. They regularly disrupt long-term forecasts and create new conditions that the designers and managers of systems need to face. The fact that trend-breakers are routine and significant means that designers have to pay attention to them.

The basic way to anticipate trend-breakers is through scenario analysis,[6] which attempts to identify possible "scenarios" or sets of coherent major developments that might affect the industry and the system to be installed. The focus is on a general qualitative rather than a detailed quantitative description of possible futures. A scenario, in this context, is a single concrete and plausible future path that the project and its environment might take. It is largely narrative and is accessible to a wide audience of decision makers and stakeholders. A set of scenarios normally features one that is largely an extension of "business as usual," complemented by others that represent different patterns of evolution, generally resulting from some disruptive events that break current trends. Box 4.5 illustrates some alternative scenarios for a particular case.

The development of scenarios implicitly identifies possible trend-breakers. A convincing narrative about a possible future has to include a rationale for the dynamics or chains of events that lead to a scenario that differs from a simple extension of the current situation. Once we focus on these trend-breakers, we can think about how plausible they might be. In box 4.5, for example, the trend-breaker leading to the "unreliable energy market" scenario would be a war in the Middle East. Based on the record, this is unfortunately quite plausible.

Weighing the relative probability of scenarios is problematic, however. Indeed, it is unlikely that the future will actually evolve according to one of the few possible scenarios. Scenario planning does not predict the future; it helps planning teams identify dynamics that can lead to

Box 4.5
Possible scenarios for power systems in Europe

Let's examine the meaning of scenarios by looking at major investments in electricity-generating plants in Spain, where three scenarios seemed relevant: business as usual, unreliable energy markets, and plentiful nuclear power.

Business as Usual

No major departures from current trends. Supplies of fuels remain reliable. Driven by steadily rising worldwide demand, fuel prices increase faster than inflation. Greater concern for global warming leads to strong incentives to invest in carbon capture. Regulations continue to be favorable to private investments and adequate returns.

Unreliable Energy Markets

Unrest in the Middle East disrupts deliveries of liquefied natural gas (LNG) and leads to much higher prices for unreliable supplies. These circumstances enable Russia to extort extraordinary prices for a guaranteed steady supply. The European Union steps into this crisis and caps price increases proposed by the suppliers of electric power, effectively taking over their assets at major discounts.

Plentiful Nuclear Power

Worldwide investments in nuclear power provide electricity reliably at a manageable price. This leaves LNG producers with excess capacity and drives their prices down substantially. Meanwhile, the switch to nuclear reduces CO_2 production and lowers the price of carbon permits. Lower costs of fuel and carbon permits make LNG power plants profitable.

Each scenario that differs from "business as usual" embodies one or more trend-breakers. For example, chaos in world energy supplies or a crash of the LNG markets could break the existing trends in power production in Europe and greatly shift production opportunities for electricity producers.

trend-breakers, which, in turn, could lead to different futures without having to go into the technicalities of quantitative assessment. The scenarios are useful markers that indicate the range of possible future developments in a way that is accessible to a large audience.

A successful scenario-planning exercise unblocks the thinking that assumes recent trends will continue. It makes participants recognize that the world could be a very different place. It should expand our vision of a range of possible futures far beyond the range associated with uncertainties in trend lines, which are simply possible extensions of the status quo. Scenario planning is a learning exercise for the planning team. It teaches them to recognize that their system may have to perform in an environment that could be quite different than the one that currently seems most likely. Scenario planning thus leads designers to consider ways to enable their projects to transition effectively to possible futures—in other words, to be flexible.

Step 4: Establish Forecast (In)accuracy

It is a professional responsibility to be realistic about our collective ability to predict accurately. To fail to do so—to pretend that a point forecast is precise—is deceitful. Being careful to place reasonable ranges on forecasts is similar to scientists placing error bars on measurements. It should be routine good practice.

Unfortunately, forecasting professionals do not discuss the accuracy of their forecasts. This is because forecasts are, in practice, "always wrong," in that there is a significant gap between anticipated and actual results. As chapter 2 illustrates, these discrepancies can easily be off by 30 percent or more over a decade. This is an uncomfortable truth for both clients, who want precision when they pay great sums for the forecasts, and the forecasters who sell their services expensively on the basis of their great skills and advanced degrees. Forecasting consultants prefer to talk about statistical diagnostic metrics, which are normally excellent.[7] However, excellent statistical diagnostics on historical data are clearly not the same as making accurate assessments of future conditions. Fitting an explanation to the past, which is what they normally do, is easier than producing useful predictions of the future.

The way to assess realistically our ability to predict is to compare past forecasts with actual outcomes. The way to judge the skill of darts players is to see how close they come to the bull's eye. Conceptually, this is a straightforward process. We compare before and after numbers, that is,

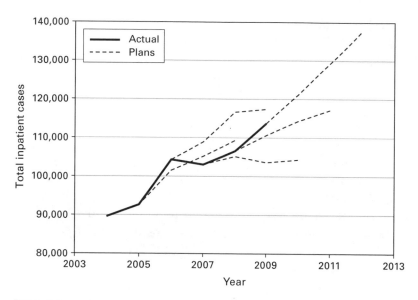

Figure 4.3
Comparisons between forecast (dotted line) and actuals (bold line) of inpatient activity
at an English hospital.
Source: Lee and Scholtes (2009).

the prediction and eventual outcome, and we note the differences. In
practice, it may require considerable effort to find earlier predictions
often made a decade or more ago. Simple comparisons, such as the one
charted in figure 4.3, can take a professional several weeks to compile
from the archival records of the relevant organization.

Comparisons between past forecasts and reality provide a good indication of how well we can hope to do for our next project. If past forecasts
were off by X percent on average over 10 years, then this figure provides
a reasonable estimate of the average inaccuracy of our current forecasts
for the next decade.

Note that the process of comparing forecasts with reality can start at
any time. Because of the difficulties in finding the results of previous
forecasts, it may be expedient to start this process early, to be able to
collect as much evidence as possible.

Step 5: Build a Dynamic Socio-Technical Model

The value provided by any large infrastructure system depends on the
demand for its services and the effectiveness and efficiency of delivery.

In financial terms, changing customer needs and competing services drive the top line; economic and operational variables such as availability, cost and productivity of labor, and cost of goods drive the bottom line. In general, a system design transforms a flow of input variables into a flow of performance variables.

To estimate the value of a system, we need models that combine both the technical realities of a system and the social, economic, and regulatory factors that will influence system performance. In short, we need socio-technical models.

The Concept

The purpose of socio-technical modeling is to represent meaningfully the dynamic transformation of technical and social performance drivers into performance indicators and relate that transformation to design parameters. Socio-technical modeling helps designers explore alternative configurations of the system and enables them to select the preferable solutions. The development of a good model should be an essential part of the design process. Poor models can lead the design process astray and destroy value, as the modeling for Boston's sewage system demonstrates (see box 4.6).

A good system model is "causal": It shows how its various elements affect each other and combines technical and social aspects. It represents both technical features (such as the way greater loads affect the size of the design) and social features (such as the way the cost of production raises prices and lowers demand for services). For example, a proper model for analyzing the development of an oil field will combine geology, structural knowledge, and operating features, on the one hand, with an understanding of the markets for oil and gas and the regulatory climate that may favor cleaner petroleum products, on the other hand.

Socio-technical models are extensions of traditional engineering models that change their nature profoundly. In contrast to our aspirations for engineering models, we have to admit at the start that we may never get the right answer from a socio-technical model. We cannot take the socio-technical system into the laboratory and test it in detail to validate the contribution made by each component. This is different than models that seek to represent technical performance accurately, such as the flow of air through a jet engine.

Socio-technical modeling is both an art and a science. It is a science in that it uses statistics to enforce mathematical rigor in developing models,

Box 4.6
Modeling for design: Boston's sewage system

The design of Boston's $6 billion sewage treatment system rested on a simple model of the interactions between the social and technical conditions. It was that:

• Sewage to be treated = function of population in service area and per person water use.

This was inadequate because it left out the crucial contribution of economics. The result was a poor design, oversized, and configured for the wrong mix of inputs. The system was unnecessarily expensive to build, and its operations were less efficient than they could have been.

An appropriate model would have recognized that the implementation of the new sewage system would greatly increase the costs to users, whose water is metered. In response, customers would reduce consumption by fixing leaks, installing low-flush toilets, and so on. As they did so over time, the price of water would rise because the requirement to pay for the fixed costs of the sewage system would force operators to raise the price per unit as usage decreased. Finally, these adjustments would change the nature of the sewage and by extension the operation of the treatment plant: A lower proportion of water in the mix, for example, affects pumping requirements and changes the chemistry of the biodegradation process.

A suitable model would have had at least the following elements:

• Sewage quantity = f(Population; Price of water changing over time, Usage changing over time)
• Sewage quality = f(Price of water changing over time)
• System costs = f(Sewage quality changing over time)
• Price of water = f(Annual costs of system, Usage changing over time)

---

and it enforces logical coherence by exposing the same encoding of all design alternatives to the same assumptions. It is an art because it requires creative simplifications of design alternatives and meaningful simplifying assumptions of the complex environments within which the system operates; it is also an art because statistics cannot prove a model.

The latter is a fundamental concept. A statistical model may fit the data well, but by itself this is not significant. A useful mantra is: "Correlation is not causality." Just because a model offers good correlation with historical observation does not mean that it displays correctly the way variables interact with each other. Box 4.7 illustrates the point. Another useful mantra is: "A good fit is not the same as a good prediction." The

Box 4.7
Correlation is not causality: Firefighters do not cause fires

---

It is obvious that the number of firefighters and fire trucks at a fire is a good predictor of the amount of fire damage. These factors correlate closely. We can confidently assume that something big is going on if we see many firefighters at a fire.

However, this model is a terrible basis for the design of a system to control fires and damage. We should definitely not conclude that sending in fewer firefighters would reduce fire damage—quite the contrary.

This is an example of spurious correlation. The correlation is real, but the causality does not exist. The actual situation is that a "big fire" causes "big damage" and triggers "dispatch many firefighters." The correlation between big damage" and "many firefighters" is there, but the latter does not cause the former.

---

fact that a model fits past data well says nothing about its predictive power. The statistical procedure selects the one model that fits the data best from among many alternatives. If there are enough models to choose from, there is likely to be one that fits well. Indeed, it is routinely possible to develop several, often contradictory, models from the same data. For example, our review identified more than a dozen doctoral dissertations and other sophisticated analyses that proposed quite different models of the use of water in Boston. Users should choose models that make sense and be aware of their limitations.

Developing Good Models

To develop a good model, we need as far as possible to incorporate our understanding of causal mechanisms. We need to combine theoretical concepts (such as the idea that price increases generally reduce demand) and practical understanding of the mechanisms at work. Some of these will be broadly applicable, for example, that the flu season increases the demands on emergency services at hospitals. Some mechanisms result from local conditions concerning regulations and other practices that induce various behaviors.

Good blending of the technical and social factors requires careful statistical analysis. At this point, it is advisable to consult a professional statistician or econometrician to help analyze causal relationships rigorously. In fact, it may be tempting, and is indeed common practice, to outsource this analysis wholesale to a consulting firm. This can be a big

Box 4.8
Designing emergency rooms for a hospital

The ER facilities at a New England hospital were consistently crowded. The hospital decided to fix this problem and called on a consultant to develop a model of ER use and propose a solution. These California consultants applied a national model to the situation and recommended a large expansion of about 25 beds.

The consultants' model was not appropriate for the hospital. They developed it without detailed input from the hospital directors and staff and thus without an understanding of the local context. If they had done so, they would have realized that in this particular situation much of the crowding of the ER was due to the local administrative practices, specifically discharging patients late in the day. This meant that ER patients needing admission to the hospital systematically crowded the ER while waiting to move on. As the hospital recognized this causal chain and moved to earlier discharges, much of the crowding in the ER disappeared.[19] The consultants' model led to wasteful overdesign.

The moral of the story is that senior decision makers and experts in the system should be part of the process for developing the system model.

---

mistake, as box 4.8 illustrates. This practice gives away important learning opportunities. It is imperative that some of the planners and designers with wider responsibility remain closely involved in modeling work, in fact lead the analysis, and try to understand and interrogate the work of the professional statisticians. If the senior designers withdraw from the analytical work, the activity can easily become a statistical exercise devoid of important context-specific insights.

Good models also capture uncertainty. They should certainly include the factors that are the drivers of design, as well as the major potential trend-breakers determined in the scenario analysis. If our models do not account for uncertainty, they will not enable designers to estimate the expected value of any design accurately due to the flaw of averages. Moreover, they will not enable them to determine the value of flexible designs, which only have value because of uncertainty.

Use of Models

Designers use planning models to test alternative designs under conditions that are as realistic and fair as possible. The practice is to expose design alternatives to the same set of assumptions about the system environment and to calculate how these simplified designs fare in the simplified worlds created. A financial cash flow analysis, for example, will encode projects using associated cost and revenue flows over time and compare alternative investments based on common assumptions about demand and discount rate, for instance. This is a fair comparison in some sense. We do not want to compare one design using a 10 percent discount rate operating under low demand to an alternative design using a 5 percent discount rate with high demand. Nor do we want to compare a project that includes headquarter overhead costs with another project that excludes them. However, the fact that the models lead to fair comparisons says nothing about their accuracy.

Accuracy of Models

It is important to understand that no model can expect to predict the performance of a complex socio-technological system accurately. The accuracy of a model's performance projection depends crucially on assumptions about causal relationships and the simplifications made to set up the model. Therefore, the results of models can only be interpreted relative to these inputs—and never in absolute terms, as in "the NPV of the project is X."

Once we accept this, it should be clear that model development cannot be delegated to a team of accountants, statisticians, or consultants, even though they may well be indispensable participants in the process. The designers and decision makers, those who are ultimately accountable for a project, have a responsibility to be involved in developing the model and its assumptions. It is surprising how often decision makers neglect this advice in practice.

Although no one can guarantee the accuracy of a model, we can perform several validation checks to increase our confidence in it. First, we must double-check all inputs, which are often not what they seem. The definitions of the data frequently differ over time, within different institutions, and when collected for other purposes. Second, we should consider how the model correlates with historical data. We can do this using the R-squared metric of goodness-of-fit, but we need to keep in mind that this statistic is unreliable as we mentioned earlier. Third, we can work on partial data, putting ourselves some years back and computing how the model, calibrated on this partial information, would have predicted the recent past; we compare model forecasts and the actuality we know. Overall, we need to be cautious. None of these approaches can prove a model's accuracy.

Application to a Major Project: Planning Hospital Capacity

This case illustrates the forecasting procedure in the context of a development plan for St. Mary's maternity hospital in England.[8] Between 2001 and 2008, the number of births delivered in the hospital increased 22 percent, from 4,479 to 5,456. The trust managing the hospital adapted the infrastructure as much as possible over the years to cope with the extra work. Finally, the trust announced plans for an extension of the maternity clinic to enable it to cope with demand over the next decade. The question is: What demand should they design for?

Step 1: Identify Key Performance Drivers

Management measures the performance of this hospital on three dimensions: the quality of its clinical outcomes, the experience of individual patients, and the financial viability of its operation. There is good evidence that health-care architecture and design affects all three objectives.[9] The most obvious drivers of hospital performance that have a direct impact on design are scale and spectrum of demand. Getting the match between capacity and demand right is a critically important determinant of the hospital's ability to provide high-quality clinical service and cover capital costs.

As an illustration of forecasting, this case study concentrates on the aggregate number of births delivered at the hospital. Once we have determined the variables in which we are interested, we need to specify the timescale within which we want to aggregate them. Annual volumes are commonly used for long-term planning. An annual timescale, however, reduces data variability considerably compared with shorter timescales, such as daily and weekly activity. Designers have to be aware of this and build in extra buffer or short-term design flexibility to allow for short-term peak demands.

The time frame chosen for the driver variables has to be commensurate with the design flexibility we want to evaluate. If we are interested in the value of short-term design flexibility, such as altering room capacity by adding an extra bed or using flexible partitioning walls, we need data on the variation in the demand at a more granular, daily, weekly, or monthly level.

In this case, the hospital is interested in longer-term capacity flexibility, that is, the design of a smaller initial building with an option for expansion. We therefore develop projections of annual demands.

Step 2: Analyze Historical Trends

We start by examining the time series of the variable of interest that we have identified in step 1, the annual number of births. The hospital had data for the previous 13 years, from 1996 to 2008. We can estimate a historical trend by fitting a regression line to these data. As figure 4.4 shows, the estimated trend is that the number of births increases by 51 per year on average.

A trend analysis has two key problems. The first is that any trend analysis is based on the assumption that the dynamics that shape the future are similar to those that shaped the past. This is not to say that the future is precisely like the past. Trend analysis does allow for unexplained year-on-year variation, as illustrated by the meandering of the past activity pattern around the trend line. The assumption is that the mechanisms that generate this variation around the trend are fairly stable over time. The mental model for the variation is the rolling of dice. The critical assumption is that the dice used to generate future variations around the trend line are the same as those used in the past; not some new and possibly loaded dice.

In practice, it is often difficult to explain the difference between a projection as an expectation around which there will be unexplained variation and a precise prediction. For example, according to the equation of the trend line, the expected number of births in year x = 2018 would be 5,595. Yet it is easy to argue against this projection. In 2008, St. Mary's already had 5,456 births. Is it credible that the number of births will only increase by 2.5 percent over the decade, when the hospital has

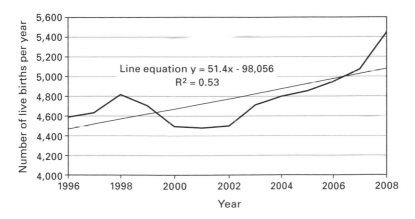

Figure 4.4
Recorded live births at St. Mary's Hospital, 1996–2008.

just experienced a 22 percent increase over 8 years? On inspection of the trend line in figure 4.4, we may argue that the 2008 record was an extreme blip above the long-term trend and is therefore inappropriate to compare with the projected activity in 2018. This argument, however, assumes that the same dice generated past variations around the trend line as future variations. The argument is wrong if something fundamental had changed—for example, if a neighboring maternity hospital that used to be the main competitor had closed—which would suggest we had introduced loaded dice.

The second problem with many trend analyses is that they are based on insufficient data. Thus, St. Mary's had only 13 years of birth data and wished to project 10 years ahead. Can we confidently predict long-term trends based on a relatively short-term record? If only limited data are immediately available, it is often worthwhile to extend the record by identifying other variables that are correlated with the variable of interest and for which more data are available. In this case, we had data on the number of births in the county from 1974 and 2008. Although St. Mary's does not deliver all births in the county and serves a population beyond the county, figure 4.5 shows an astounding correlation between the number of births in the county and in the hospital. This makes it plausible to use the county data to develop a longer-term trend in births at St. Mary's.

During the period 1996–2008, the number of births at St. Mary's averaged 75 percent of those in the county. In view of the close correlation

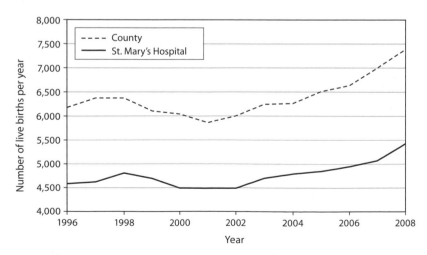

Figure 4.5
Recorded live births at St. Mary's Hospital and in the county as a whole, 1996–2008.

between the county and hospital volume, it makes sense to use the county data to estimate the unknown earlier historical births at St. Mary's by multiplying the county data by 75 percent. Because the county's birth figures go back to 1974, they provide 34 years of simulated data for St. Mary's, as figure 4.6 shows.

Note that figure 4.6 contains two trend lines. One fits the original birth data from 1996–2008 and is already contained in figure 4.4. The other fits the birth data estimated for 1974–2008. These trend lines are quite different. The 2018 projection is 5,220 based on the longer time series, much lower than the 5,595 estimate based of the first trend line. This longer-term trend is arguably even less credible for the hospital management and staff given the 5,456 births in 2008.

It is worth dwelling a bit on the issue that the low volumes forecast by the trend lines seemed unacceptable in the light of the most recent data. In fact, the trend lines do a bad job at predicting the number of births in the most recent years. So how good can they be at predicting the future?

This raises an issue inherent in the process of trend line fitting. The mathematical analysis gives equal importance to all past data; it does not give any special consideration to the more recent, and plausibly more relevant, record. There are ways of producing trends that give the more recent data more weight, such as moving averages or exponential

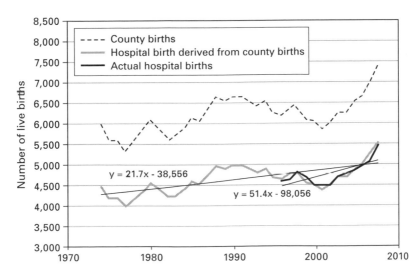

Figure 4.6
Historical birth volume at St. Mary's Hospital generated from county data.

smoothing methods.[10] An alternative and extreme position is to argue that:

• the most recent observation, in this case the number of births in 2008, is most important in determining future birth numbers; and

• all that the past tells us is how birth numbers change from year to year.

This position subtly changes the model of the process that generates the forecast number of births.

The original model assumed that there is a long-term trend in birth volumes represented by the trend line. Year-on-year deviations from that line arise because every year nature draws some random noise, akin to rolling dice, and adds it to the point on the trend line. The expected volume, and our best prediction, in any one year is therefore the point on the trend line.

The alternative model zooms in on the most recent observation, the 2008 number of births, and assumes that it is our anchor point for the future. To get predictions for future years, it simply extends average year-on-year growth from this anchor figure. Any year-on-year deviation from this projection arises because nature draws some random noise and adds it to the annual growth. The key point is that this model anchors projections on the most recent observation; any fluke that may have led to this most recent observation is carried over to all future projections.[11]

Let's see how the alternative model works for our data. To estimate the long-term growth, we consider annual changes in birth volumes since 1974. These data show an average increase of $(5,456–4,463)/34 = 29.1$ births per year. If there is reason to believe that the expected annual growth changes over time, we might perform a trend line analysis of the year-on-year change in birth numbers.

In summary, there are many ways of producing projections based on trends. Figure 4.7 shows that different approaches can give rise to quite different projections.

Step 3: Identifying Trend-Breakers

Recall that in step 1 we identified the total number of births as a critical uncertainty for the hospital design. Step 2 then gave us a first idea of historical trends. However, the future may not proceed as the past. To identify potential trend-breakers in the number of births, we need to understand the key factors that affect this number. As figure 4.5 shows, the number of births in the hospital has historically been close to 75

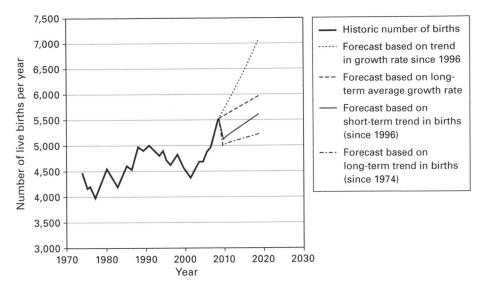

Figure 4.7
Forecast birth numbers in St. Mary's Hospital based on various trend analyses.

percent of those in the county. So in our case, it is sensible to focus on the key drivers for the aggregate number of births in the county, which are the number of fertile women in the county and their birth rate. As a first approximation, we can model the number of births in the county in year t as[12]

Births in the county at time t = Fertility rate (t) x Number of women aged 15–44 (t).

Figure 4.8 shows the historical evolution of fertility rates and number of fertile women in the county.

The fertility pattern in the county varied considerably between 1974 and 2008. During the 1970s and 1980s, the fertility rate declined in general but with considerable year-on-year variation. During the 1990s, it declined rather steadily and hit a record low in 2001. Since then, the fertility pattern seems to have reversed, and the fertility rate has been constantly increasing. This is an example of a potential trend-breaker. It is important to understand why this has happened to judge whether this trend will continue.

A careful analysis of the causes for potential trend-breakers can lead to useful possible future scenarios with clearly defined assumptions. For instance, Tromans and colleagues from the UK Office for National Statistics have explored the factors behind increasing fertility rates

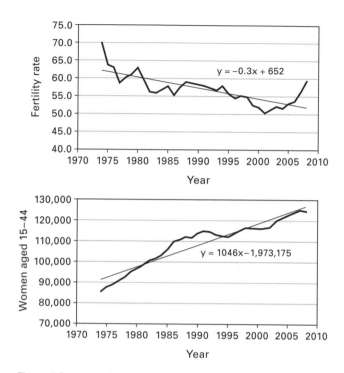

Figure 4.8
Number and trend of births per year per 1,000 women aged 15–44 (top) and number and trend of women aged 15–44 (bottom) between 1974 and 2008.

across England since 2001.[13] They argue that the rise is mainly due to growth in the foreign-born population, which exhibits higher fertility rates across the board, and to an increased fertility rate in two age groups of UK-born women, those aged 20–24 and those aged 35–44. Two sample scenarios for St. Mary's might thus be:

• *Scenario 1* The size of the foreign-born population will increase further, and fertility rates among UK-born women continue to be high. For this scenario, it would seem appropriate to estimate the fertility rate trend based on the data since 2001.

• *Scenario 2* The size of the foreign-born population and fertility rates among UK-born women will reverse to the long-term trend. For this scenario, the trend should be estimated using the full-time series since 1974.

Figure 4.9 shows the different growth trends of fertility rates and birth number projections associated with these two scenarios.

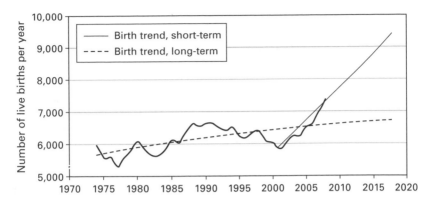

Figure 4.9
Birth trends for long- and short-term scenarios.

Step 4: Establish Forecast (In)accuracy

Having identified the critical variable in step 1, investigated appropriate historical data in step 2, and discussed trend-breakers on the basis of a refined model in step 3, we may (or may not) be able to forecast the future. The goal is to produce a graph similar to figure 4.3, which illustrates how well we have forecast in the past. Unfortunately, although most organizations keep good records of past performance data, few keep records of past forecasts. In the absence of real forecasts, it is useful to simulate how a forecasting procedure would have fared had it been applied in the past.

We illustrate this process using the number of births in the county, for which we have data from 1974 to 2008. Recall that, based on figure 4.5, we can derive a forecast of the number of births in the hospital by multiplying the number of births in the county by 0.75, the stable long-term fraction of county births delivered in the hospital. To simulate past forecasts of county birth numbers, we begin in 1980 and produce a 10-year forecast. For illustration purposes, we perform the forecast in a simplistic way: We calculate the average annual rate of change in the number of women aged 15 to 44 and in the fertility rate over the data from 1974 to 1980; we use these to produce 10-year projections of the number of fertile women and their fertility and then multiply these two forecasts onto one another to obtain a forecast of the number of births in the county from 1980 to 1990 based on 1974–1980 data. We repeat this process with other base years, 1981, 1982, and so on, using the historical data that we have had in the base year as the basis for the forecast. Figure 4.10 shows five such forecasts, as dotted lines, against the actual birth

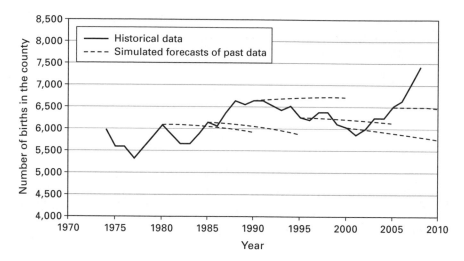

Figure 4.10
Simulated 10-year forecasts for county birth numbers.

numbers. This is the type of figure we want to produce in step 4. Compare figure 4.10 with figure 4.3, which is based on real rather than simulated forecasts: They are not dissimilar, and both are good illustrations of the inaccuracy of the forecasting process.

Of course we can and should experiment with different forecasting procedures to try to improve the accuracy of historical forecasts. We might argue, for example, that short-term trends are more adequate than the long-term trends accounted for in the model in figure 4.10. This is easily done by considering average growth over a fixed period (e.g., over the past 7 years) rather than over the period of the available data. However, this does not improve the accuracy in this case, as figure 4.11 illustrates.

Step 5: Building a Dynamic Model
The final step of the forecasting process requires us to build a dynamic forecasting model, that is, a spreadsheet module that produces a range of forecasts drawn from a sensible distribution. A reasonable way to do this is to apply the forecasting procedure we used in step 4 forward from the current year—in our case, from 2008. This will produce a line projection. To address uncertainty, we need to add a random error to this line. The question is, what is a sensible range and distribution of errors?

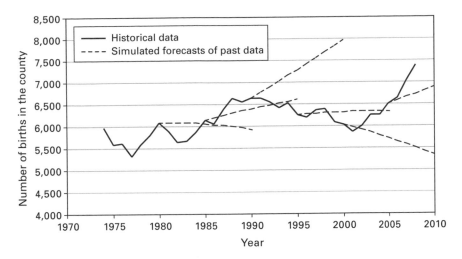

Figure 4.11
Simulated 10-year forecasts for county birth numbers based on preceding seven years of data.

Note that we have actually generated a sample of possible errors with our simulated forecasting procedure in step 4. Why not use those? Figure 4.12 shows five error paths. Each corresponds to successive errors made in a single projection (i.e., the error in year 1, year 2, up to year 10). Figure 4.12 illustrates that the range of errors in longer-term forecasts is larger, which is to be expected because it is more difficult to forecast further into the future.

Our simulation model of the future now consists of producing a line projection, as we explained above, and then choosing randomly one of the error paths from our simulation in step 4 and adding it to the line projection. We can repeat this, obtain another random error path, add this to the line projection, and so on. This allows us to simulate different future paths in a way that is consistent with the data. Figure 4.13 shows a range of such paths.

Take Away

This chapter outlines a forecasting process that takes uncertainty into account. This is an indispensable prerequisite for a systematic appraisal of flexible designs. Flexibility reveals its true value only in the face of uncertainty. The procedure outlined here will lead clients and designers to appreciate the level of uncertainty in their projections. But it does

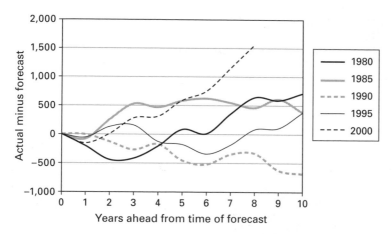

Figure 4.12
Deviation of forecast from actual over time for five simulated historical forecasts.

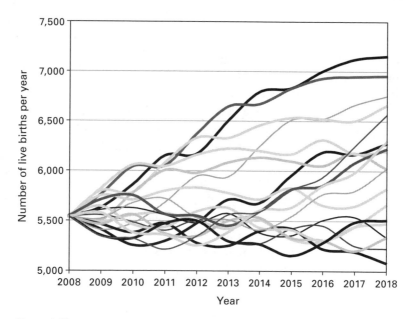

Figure 4.13
Dynamic forecasts of hospital births. Nineteen scenarios of birth number evolution, based on a projection in 2008, adjusted by 1 of 19 forecast error sequences obtained by simulating 10-year birth projections annually between 1980 and 1998.

more than this. It allows us to replace the prevalent single-number projections, epitomized by the trend line graph, with a spreadsheet model that generates many different paths of the future environment in which a system may have to operate. These simulated futures are all equally likely. There is not "one future" against which to design the system but many possible paths from which nature will select. The design challenge is to produce a system that can cope with several of these futures.

Appreciating the uncertainty in projections can be uncomfortable, and therefore it is often avoided. It is much easier to make a case for an investment to a company's investment board, a bank, or a government agency if we can explain the economics based on single number projections, as if we knew what the future brings. However, this is likely to lead to ineffective and inefficient systems, as we demonstrate in part I of this book. The dive into the cold water of uncertainty requires courage; it is a challenge for clients, financiers, and designers alike.

Once we recognize uncertainty, we can approach design with robustness or flexibility. Robustness is the world in which traditional engineers feel at home. It requires a reasonable overdesign of the system so that it functions well under all scenarios. However, this approach commits all capital upfront and is thus very costly because only one of the futures will occur. In many circumstances, it is smarter to build enough flexibility in the system to allow the system operator to adapt it to changing circumstances. We explain how to do this systematically in the remainder of this book.

5   Phase 2: Identifying Candidate Flexibilities

Everything should be made as simple as possible, but no simpler.[1]
—Attributed to Albert Einstein

This chapter describes ways to identify the most valuable kinds of flexibility for a system. This is an important topic. Having accepted that the future is uncertain, and so concluded that it would be good to have a flexible design, the question is: How should we implement flexibility? We need to know: What parts of the system should be flexible? How flexible should they be?

It is not obvious which flexibilities will add the most value to a project. The answer depends on many interacting factors, such as:

• *The nature of the system*   The flexibility needed to make an automobile plant responsive to changes in the market demand for different types of vehicles is different than the kind of flexibility needed to develop a copper mine. The automobile factory might benefit from having multipurpose robots; the copper company might choose to access the ore deposit through an open pit rather than tunnels.

• *The kinds of uncertainties*   Are these mostly on the supply side, such as the state of available technology? Or are they on the demand side? For example, are we mostly concerned with the size of the market, as we might be for a new product, or with the price we might get for the product, as when we develop a system for a commodity like oil? Are we interested in the flexibility to expand or contract easily, or to redeploy capabilities and equipment to parts of an oil field that might only be profitable at high prices?

• *The intensity of uncertainties*   Is the system well established and slow to evolve, like a rail transit network in a big city, for example? Or are we dealing with a situation that is fluid and subject to rapid changes, such as the air transport industry in a deregulated environment? Are we interested in long-term options, such as the ability to double-deck a bridge? Or should we focus on short-life design, the kind that enables us to reconfigure airport terminals for different types of user or aircraft?

• *The cost of implementing flexibility*   How much might it cost to initiate a form of flexibility? How much effort will be needed to take advantage of flexibility later on, when it might be useful?

In general, we can incorporate many different kinds of flexibility into the design of a system. A number of these will improve value. But which are likely to add the most value to the system? Which should we focus on?

This chapter provides a practical guide to best current practice for determining the better kinds of flexibility to build into a system. The recommended approach is to use screening models to examine the design space rapidly, and thus to identify a short list of attractive design concepts that we can explore in depth. We begin this chapter by presenting the three ways to develop screening models: bottom-up, simulator, and top-down. Each of these can be useful. The choice between them depends on the details and the nature the problem to be addressed. We end with a case study on the use of screening models to identify desirable flexibilities in the configuration of automobile factories. For further guidance on how to identify interesting possible flexibilities in specific fields, please consult the wide variety of case studies posted on the Web site for this book (http://mitpress.mit.edu/flexibility). The field is developing rapidly.

Concept of Screening Model

The recommended approach is to use screening models for preliminary identification of the most desirable candidate flexibilities. Screening models are simple, understandable representations of the performance of the system or project under development. The chapter first defines the concept of screening models and shows why we need them. It then presents in detail the main approaches to developing screening models: bottom-up descriptions of detail, "simulator" models, and top-down conceptual representations. Finally, it explores how we should use screening models to identify desirable candidate flexibilities.[2]

A screening model is a simple representation of a system that can be used for a quick analysis of design performance. For example, the screening model used to describe the economics of alternative designs for oil platforms might take a couple of minutes to estimate the value of a specific design under a single scenario. In contrast, the full oil-and-gas model used for the final design might take a day or more to do a complete analysis of the same case. The complete model combines detailed descriptions of platform design, the behavior of the flows in the oil field, and the economics over time. Each submodel will have been developed by different expert groups and will not have been fully integrated with the others. Such complex models are common in the design of technological systems. Obviously, we cannot use models that take a day or more to examine a single design for a single scenario to examine and compare several designs over thousands of scenarios—there are not enough days in the year. By contrast, screening models allow us to look at thousands of alternatives very quickly, typically in less than a day. Speed is necessarily a primary characteristic of a screening model. Speed enables a screening model to fulfill its function of providing rough estimates of how many different designs perform over many different scenarios.

We need fast models so that we can identify as reliably as possible the best opportunities for creating value. Screening models allow us to explore the extraordinarily large number of combinations of possible designs and paths of evolution of uncertainties over time (see box 5.1).[3] We need efficient means for searching for good designs to complement our intuition or experience with what worked in the past. Intuition and creativity are important assets but can often lead us astray, especially about complex combinations that we have not experienced. Researchers have repeatedly shown that our intuition on probabilistic effects is particularly weak. Standard psychological tests, of the kind used in introductory classes on behavioral science, demonstrate our collective difficulty in dealing with uncertainty. People rapidly fix their thoughts rather arbitrarily on notions that often turn out to be wrong.[4] Likewise, experience about what has functioned elsewhere, under other circumstances, is not necessarily a solid guide to what may be best in a different, specific situation. As designers, we need rigorous analytic processes to complement our valuable intuition and experience, and to help us examine the many alternatives. Because of the large number of possibilities, we need fast models to do this job.

Counterintuitively, these fast models may not be readily available. Technical organizations are likely to have large, complex representations of their operations, assembled from a collection of detailed models of individual processes. An oil company, for example, may have an oil and gas model that combines large interlocking computer programs connecting design, production variations over time, logistics, and economics.

Box 5.1
Large number of possible combinations

The number of possible designs for any system rapidly becomes huge. Consider the simple problem of contracting for warehouse space at 10 locations, which can either be "large" or "small." This amounts to 2^{10} or about 1,000 possible configurations. Suppose we can change our contracts in three periods and can decide whether we should have a warehouse in a location and whether it should be large or small. Then the total number of configurations grows to $1,000^3$ or 1 billion.

The number of possible development paths for some uncertainty over time is likewise very large. Consider the development of an open pit copper mine over 20 years. Its overall value depends on variations of the price for copper over time. If copper sells for a high price early in the development, this will lead to higher present values than if high prices occur later. Furthermore, different prices will lead to different development strategies or mine plans: When prices are higher, it is rational to focus on immediate exploitation of richer deposits to take advantage of immediate opportunities. When prices are lower, it may be better to concentrate on removing overburden and to position the mine for rapid access to deposits later on when prices are higher. Therefore, we should consider the effect of possible sequences of prices. If we simplistically think that prices can either rise or fall some amount (say 10 percent) from year to year, then the total number of possible distinct price paths is 2 to the power of 20 = (2^{20}) or more than 1 million!

In general, it takes several hours to process a development strategy for a mine for a single sequence of prices. This is because the mine plan is the solution to a very large optimization problem. Analysts divide the volume of the mine into cubes, perhaps 20 meters on a side. There may be 100,000 of these blocks in a major mine, each of which has an estimated ore content. The mine plan thus defines the most profitable sequence of removing these blocks, subject to physical constraints (blocks on top have to come out first) and the capacity constraints of the mining equipment. It is not practical to run this optimization thousands of times. We thus need some simpler model to enable us to consider the effect of the range of possible price sequences over time.

Such industrial models may easily take many hours to run to obtain a result for a single set of possibilities. These models are not appropriate for analyzing hundreds of scenarios. The creation of a useful fast model may require significant effort.

Screening models should also be simple and understandable. A simple structure makes it easier for planners and analysts to see how the individual parts of a system affect each other and so appreciate how various decisions may affect overall system performance. An understandable model helps build confidence in the analysis process. When decision makers can see how and why a model works, they are more likely to accept both it and the resulting analysis. By contrast, a complex model whose interactions are unclear may appear untrustworthy—often rightly so. Moreover, large industrial models may be poorly documented and difficult to understand, especially once the experts who created them have left the organization. Such models are unlikely to enable us to enhance our intuitive thinking.[5] Conveniently, simple models are also fast.

Screening models are thus both qualitatively and quantitatively different from large complex models. They should be easier to understand from an overall perspective so that system planners and designers can develop confidence in them and trust their guidance. They should be simple so that they can rapidly scan the wide range of possible designs and identify configurations that are worth examining in detail.

We use screening models to search the vast space of system configurations and possible futures and determine a short list of attractive possibilities for configuring and operating complex systems. We use them to explore quickly the range of design possibilities and uncertainty scenarios. If we think of all the possibilities as an unexplored continent, screening models enable us to fly over the terrain at high altitude and identify interesting peaks. Once we have found some high points, we can investigate their possibilities in detail. The name for screening models describes their function: They screen or filter out interesting possibilities.

Screening models complement the complete, complex models needed to design a system in detail. They are not substitutes, and they are most useful in the early stages of the design process, when we are trying to decide the overall features of a system or subsystem. They help us investigate what kinds of flexibilities we might best incorporate into the design, how we might organize the system into modules, and how we

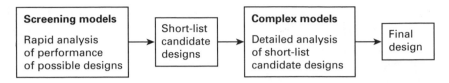

Figure 5.1
Screening models precede and complement detailed design models.

might phase development over time. Screening models help us understand and define interesting configurations or architectures for the system. Once we have used screening models to identify interesting design possibilities, we have to examine them in detail with more complete, detailed analyses. Figure 5.1 shows this sequence, from "coarse" to "fine" models.

Best practice uses screening and detailed models together. In the first stages of a project, we should use screening models more to develop a rough understanding of the values of a broad range of designs. Once we have narrowed down the range of design alternatives, we use complex models to test and validate the general results of the screening models. Later in the design process, detailed models refine and modify the design configurations suggested by screening models. We may also use screening models to explore particularly interesting features that emerge in the design process. Screening models have a more prominent role at the start, and complete detailed models dominate as the project nears final design.

The Development of Screening Models

A good screening model should have two principal features: It must be fast so that it can provide a quick way to look at designs under many different scenarios; and it should rank alternatives reasonably correctly, so that the short list represents excellent candidates for detailed examination. To fulfill its function of identifying good candidate designs, it must identify the most preferable through comparison with others.[6]

Note that screening models do not have to be accurate, in the sense that they measure the performance of a design precisely and correctly. That is not their function. Their role is to define a short list of possibilities. We can use detailed models to refine design specifications and provide more exact assessments of design performance. We need to recognize and accept that screening models will not be precise—they are stripped-down representations of system performance and are necessarily approximate.

The distinction between the requirements for screening models and detailed models is important. Professionals who normally work with detailed design models, and who have probably contributed to their refinement and precision, naturally resist the use of simple models. They are likely to consider screening models poor substitutes for the "real thing." They will want to "improve" screening models by making them more like detailed models. But we must resist the tendency to complicate screening models unnecessarily. Screening models are not substitutes for fully detailed models; they are complementary assistants or scouts, looking, as it were, for good opportunities for the troops to exploit with a detailed design.

Screening models do not have to be precise. They serve their purpose by being fast and enabling designers to search the entire solution space quickly. We can accept their lack of great precision because we know we will follow up their indication of good designs with detailed analysis.

Types of Screening Model

Screening models belong to three generic types:

• *Bottom-up* Simplified versions of the detailed descriptions of the system, built up from understanding of the parts;

• *Simulator* Mimicking detailed descriptions of the system; and

• *Top-down* Conceptual representations of the overall pattern of system behavior.

These categories indicate the different paths that analysts take to arrive at the goal of creating simple models they can use to understand a system and define good opportunities for flexibilities that will increase value.

In practice, screening models for a particular problem may be used in combination. A top-down process might describe the interactions between major modules of the screening model of a system, and a bottom-up process might represent the working of particular modules. For example, analysts might set up a top-down model to describe the overall interactions between the emergency and in-patient facilities at a hospital, the way patients transfer out of emergency care and get admitted to regular wards, and how admissions patterns, health-care policies, and operational constraints determine the availability of beds in the hospital.

Within this context, a bottom-up model might detail the operation of emergency facilities, for example, how doctors triage patients and move them on to the next stages of care. Analysts can construct a useful screening model, providing the means for rapid exploration of alternatives, in many ways.

Bottom-Up Screening Models

Bottom-up screening models are more common because they are easier to develop. They build directly on the detailed technical knowledge of the professional teams responsible for the development of a system. The challenge is to simplify meaningfully existing knowledge of system complexities.

A major obstacle in creating a screening model from the bottom up is the possible difficulty of meshing the detailed knowledge about the parts of a system into a coherent whole. We may face this problem when the system is an assembly of unrelated elements, or when it is unprecedented—such as the rehabilitation of a large area of land for the development of the London Olympics 2012 site. In such cases, we have to create an overall model for the complete system. The difficulty then is that the different specialists associated with the project will have computer models based on different kinds of data, organized at different levels of detail, and intended for different purposes. Often these specialists have not worked together closely, so they have not coordinated their efforts. In this case, the development of a suitable screening model for the system may require a significant effort—which is the case for developing integrated models of forest fires (described in box 5.2).

The better and usual situation is that the professionals already have an integrated set of models to describe the system. A project team will typically have been working on similar projects for some time and developed ways to coordinate these models into a functioning suite (see box 5.3). These models bring together knowledge about:

• Technical possibilities that engineers and planners can implement;

• Environmental factors, for example, displaying the effects of climate or the dynamics of flows over time; and

• The economic or other benefits that a project may generate.

Such suites of models provide a good basis for the development of a suitable screening model.

Box 5.2
Creating a system model: Forest fire abatement

Forest fires are a major concern in many parts of the world, especially in large timber-growing regions such as Australia, Indonesia, Portugal, Russia, and the Western United States. "At the beginning of the 21st century, the... Earth ... annually experienced a reported 7 to 8 million fires with 70,000 to 80,000 fire deaths."[14] Forest fires also often destroy towns and have great regional impacts. These impacts motivate the search for effective strategies to reduce the damaging effects of forest fires.

The design of a regional fire abatement program needs to integrate several major components, such as:

- The climate and meteorological patterns of wet and dry seasons;

- Topological descriptions of the terrain, which affects how fires spread;

- Descriptions of the composition of the forest, as different species burn differently;

- Detailed models of the dynamics of fire fronts advancing through a forest; and

- The effects of firebreaks and other actions to limit the spread of fires once started.

Moreover, we have to model all these elements over time as forests mature and ground cover changes.

Many experts have developed these elements of the system, typically in different organizations, often in different countries, and certainly in different ways. Integrating them into a coherent whole requires a special effort.[15]

A useful screening model for this system will have to simplify many of these models. For example, much of the detail about how a fire front advances will not be useful in defining how we should develop the most effective regional fire abatement strategy. A simplistic fixed advancement rate may well be preferable at this high level of decision making. Note, however, that using an average instead of a distribution of possibilities could lead us to the flaw of averages unless we are careful!

Box 5.3
Examples of integrated system models

River Basin Development

Planners for the development of river basins use models that integrate:

• A "river flow model" that describes the downstream progress of water both naturally and as released by dams and irrigation channels;

• A "hydrologic model" of the varying patterns of rainfall, snow melt, and run-off; and

• An "economic model" to calculate and optimize the cost and net value of flood control, agriculture, and electric power that the dams and other projects generate.

Each of these can be more or less detailed depending on whether models consider detailed weekly or monthly flows or work with average flows over a season or a year.

Oil and Gas

Developers of oil fields use "oil-and-gas" models that combine:

• A "facilities model" that defines the sizes and details of the many elements needed to pump the crude from the field; separate the oil, gas, and water; compress the gas; store the products for transshipment; and so on;

• A "field model" that describes the dynamics of the flow of oil, gas, and water through the ground to the wells in response to both natural pressures and pressure created by injection wells; and

• An "economic model" that translates the cash flows of capital expenses, operating costs, and revenues into net present values and other measures.

Each of these models is highly complex. They also need to pass information back and forth between each other. A change in pumping capacity, for example, can change the flow in the oil field, which in turn affects the pumping capacity that the operators might need and use. Because of this complexity, the definition of the best design for the development of an oil field for a single scenario may take a day or more. Hence, there is a need for simplified models for the screening process.

The development of a bottom-up screening model is relatively easy if we can start from an existing model of the system. We can then implement it without much special effort. The idea is to reduce the complexity of inputs, simplify the description of the problem, and reduce the time it takes to do an analysis. For example, analysts can replace complex input dynamics with time averages or simple linear trends. They can also aggregate periods of analysis, working with years rather than weeks or months. When they apply such simplifications, they cut down the size of the problem by reducing the number of variables (by more than 90 percent if we move from monthly to annual data) and the number of associated constraints. Because the time to carry out an analysis is often roughly proportionate to the square of the number of variables, these simplifications speed up the analysis considerably.[7] Moreover, the smaller model is much easier to specify and implement, which often is a critical consideration for analysts.

When simplifying a complex model, we must not throw the baby out with the bathwater, so to speak, and omit detail that may be crucial to the value of a system. The loss of critical information typically seems to result from either individual or institutional lack of attention. Standard procedure in one oil company illustrates the point. The geologists in this corporation have detailed models of the distribution of uncertainty associated with the production of a field. Yet when their information passes to the design departments, it reduces to a single number, such as the P_{50} or median value. Because the amount of oil in a field is a crucial driver of value, elimination of information about its uncertainty is clearly an excessive simplification.

Simplified bottom-up models are common. Overall, the ease of creating them, combined with the effectiveness of this approach, accounts for their widespread use. See box 5.4 for an example.

Simulator Models[8]

Simulator models are an excellent way to develop screening models. A simulator model ignores the inner workings of a process, focusing instead on reproducing the outputs of an existing complex model from its inputs. It treats the technical model as a black box and simply mimics the results of that process. The idea is to represent the output of a suite of models representing a system through a formula or simpler set of models. In contrast to simplified models, which change the inputs to the integrated model, the simulator approach focuses on the outputs.

Box 5.4
Simplified bottom-up screening model: River basin in China

Working with a Chinese investment bank, the research team analyzed the development of a major river in southwest China, whose potential for hydropower, flood control, and irrigation was undeveloped. The plans included a set of possible dams, diversion tunnels, and irrigation projects. The design of the overall system had to specify both the size of each project (such as the height of each dam) and the sequence and timing of the development.

The team developed a screening model for the river basin, a simplified version of the detailed models designers were using to specify the details of the development. It worked with average annual flows rather than stochastic seasonal flows, and it eliminated a number of possible developments from consideration because previous analyses indicated they were not desirable. These simplifications drastically reduced the size of the stochastic, mixed-integer programming problem. The simple version had less than 1/10th the number of constraints and variables than the complete problem. The screening model could develop an answer in a few hours on a laptop computer, rather than the several days needed to run the complete suite of detailed models on a larger machine.[16]

The development of a simulator model is largely a statistical exercise. The idea is to fit a simple model with a few parameters to the output of a complex model with many variables. This is akin to curve fitting, where a set of outputs corresponding to a set of inputs are observed to identify a simple curve that reproduces the outputs from the inputs. It is a two-step process:

1. Identifying the most important variables (hopefully few); and

2. Fitting a simple functional form to a series of model responses.

There are two ways to develop simulator models: direct and indirect. The direct approach is a process for replicating the output of the complicated, detailed model. It ignores the interior mechanics of the system and focuses on mimicking the overall results. The indirect approach is more subtle. It develops new, approximate models for the outputs of the components of the complex model, and it accepts them if they produce overall responses that are sufficiently similar to the results of the complex models. Let's look at each in turn in more detail.

The Direct Approach

1. Select a set of system requirements and parameters the system must respect, define the range of values that these might have in different situations, and specify a limited number of values for each of these parameters over these ranges. For example, the price of oil is an important parameter for determining the value of an oil field. The analyst might choose to assume that it will vary between $20 and $160 per barrel and fix on considering specific values $20 apart, that is, prices of $20, $40, $60, up to $160 per barrel.

2. Run the integrated system model to find the output of the system for all, or a large number of, the possible combinations of the chosen values of the input parameters. For the oil field example, this would be for all the combinations of the price of oil, the estimated size of the reservoir, the capacity of the pumping facilities, and so on. (This is a "full factorial" exploration of the response of the system.) The result, for instance, would be the NPV for each combination of parameters.

3. Develop an equation that most closely correlates the input parameters to the results. This step involves a standard statistical analysis.

4. The resulting equation is the simulator screening model.

Box 5.5 presents the results for an application in the configuration of oil platforms.

The direct approach to developing a simulator model is particularly attractive because of its ease of development and use. The development process does not require detailed knowledge about the inner elements of the system. Because the statistical analysis is elementary, all we need to develop a simulator model is the ability to run the integrated model and the time to do so for enough combinations to provide a satisfactory representation of the overall response of the model. (What constitutes "enough" depends on whether the values generated change rapidly or smoothly, and on the cost of running the models through more combinations of the parameters.) The simulator could not be simpler to use. It is an equation with a limited number of parameters—something that a computer can calculate in no time.

The drawback of the direct approach is that it offers no insight into the inner workings of the system. The simulator may tell you, for example, how the cost of a system varies with its size—but the equation gives no clue as to *why* costs vary as described. The simulator does not allow you to look into the system—the final equation is all there is.

Box 5.5
Simulator direct approach: Oil field development

Designers for deep-water oil platforms use oil-and-gas models to define the details of a prospective platform. These models combine detailed descriptions of the design of the platform, the behavior of the flows in the oil field, and the economics over time. In practice, it may take a day or more to use an oil-and-gas model to do a complete analysis of a single design for a specified set of parameters.

It is straightforward to develop a simulator that reproduces the overall value of a platform calculated by the detailed suite of models. First, we assume the principal drivers of value. For example, analysts might agree that these are the size of the field, the capacity of the platform, and the price of oil. We then run the detailed model for combinations of these value drivers. For each combination, we get a value for the system. Finally, we do a statistical analysis to find the best fit for an equation, such as:

Value = a(size of field) + b(capacity of platform) + c(price of oil) + e

If the value function appears nonlinear, we might prefer an equation such as

Value = e(size of field)^a(capacity of platform)^b(price of oil)^c

In any case, the procedure is similar.

A simulator model has the valuable advantage of allowing the analyst to explore the overall behavior of the system and confirm that it makes sense. For example, we can use a simulator model to see whether the detailed complex model correctly shows reasonable and acceptable economies of scale. We need to verify the nature of such crucial overall features of a system, which are not visible from close-up examination of particular points. This ability to examine overall features is a particularly important capability. Indeed, it can easily happen that the integrated model of the system gives wrong, even impossible answers—but nobody notices because they see only the detail, the trees, and not the forest. The institutional structure may create this kind of myopia, which can occur when the area specialists developing the detailed model focus on the components within their particular competence and do not get to appreciate the overall behavior of the integrated model. The experts in a limited area are responsible for only a small part of the model and may never get to see the big picture. Similarly, the analysts working with the entire complex model may be too busy using it in a project to explore its individual features. In any case, it is desirable to validate the performance of the overall complex model of the system, and a simulator model provides an easy means to do so.

Errors in the overall model may occur when its components embed assumptions appropriate to a limited context but not to the overall system. The possibility that the integrated model may have fundamental flaws is counterintuitive. Analysts often assume that an integrated model must be accurate because it combines the detailed knowledge of specialists. It does, of course, but the integrated model may lack appreciation of the overall behavior of the system (see box 5.6). A simulator model may then provide a valuable check on the modeling process and may point out the need to revise the complex detailed model of the system.

Indirect Approach

Whereas the direct approach focuses on the overall outputs of a system, the indirect approach looks first at the outputs of system parts and uses these to build a representation of the entire system. The indirect approach requires much more effort than the direct approach. It also requires substantial understanding of the technical, scientific, and social elements involved in the system or project, which means its results may provide a much deeper understanding of performance.

The essence of the indirect approach is to develop and use simple models that we can substitute for the complex models used for the detailed design. These new simple models are not simplified versions of the complex models, developed by dropping out details or linearizing nonlinear relationships. Analysts develop these models from higher level considerations (such as the requirement for mass balance) and adjust their parameters so that the simple models replicate satisfactorily the behavior of the detailed models used in design. Box 5.7 explains this by example.

Top-Down Screening Models

Top-down models provide an overall view of the system. They show how its major parts influence and interact with each other over time. A top-down model can usefully reveal complex interactions. In a hospital, for example, the construction of new wards might reduce financial losses—save money—by reducing the way emergency surges (e.g., during the flu season) force the cancellation of elective procedures and thereby

Box 5.6
Simulator check on integrated models: Oil field development

Simulator models can reveal inconsistencies in complex, detailed models. Working on the design of a major, multibillion dollar project, a research team used the simulator model to determine that the integrated model the designers were using had fundamental flaws and led to major and costly design errors.

The team examined the simulator to see how the detailed integrated model generated overall characteristics of the system. In particular, it looked at how the optimal cost of the system depended on its size or so-called "cost function." Economists routinely approximate this relationship with the equation:

Cost of System = A (Capacity of System)^B

If B is less than 1, then the costs do not grow as quickly as size, and we talk of economies of scale. (If we increase capacity by 1 percent, cost increases by B percent.) B is the "economies of scale factor." The lower B is, the greater the economies of scale. Both experience and theory indicate that the lower limit on the economies of scale factor B is about 0.6, as indicated in appendix C. Economies of scale mean that the average cost per unit of production capacity decreases substantially as we build larger plants. Economies of scale are thus crucial factors because they drive designers to create the largest economically reasonable facilities.

When the team looked at the response model for the system, it found that the detailed model implied that the economies of scale factor for the system was about 0.3. This looked miraculous—and was indeed too good to be true. It turned out that there was an error in the complex model. The actual economies of scale were much less. The consequence was that the design team, led astray by their integrated model with its implied tremendous economies of scale, was systematically overbuilding the size of their facilities.

The cause of the problem was that one of the components of the model assumed that a particularly important piece of the kit only came in one size. This size of kit was normal for a specific, very large, overall system, but it was inappropriate for smaller versions of the system. The assumption that this element only came in one size led to excessively high total costs for smaller systems and thus the implied enormously impossible economies of scale.

Box 5.7
Simulator indirect approach: Oil field development

The research team referred to in box 5.6 also used the indirect approach to develop a simulator model. The idea was to create simple models of the component parts of the oil-and-gas model. These distinct submodels corresponded to each of the major elements in the system: the reservoir, the facility for exploiting the field, and the project economics. Here we illustrate the procedure using the example of the reservoir submodel.

The basis for the simulator submodel of reservoir behavior over time was a simplified representation based on a material balance equation assuming homogeneity of the reservoir. To achieve a suitable fit with reality, in this case represented by a mature oil field for which 15 years of data were available, the analysis examined various correction coefficients to adjust the model for the flow of water and gas through the reservoir. The preferred model was the one that minimized cumulative error in terms of use of water injection and the resulting production of oil and gas. As figure 5.2 indicates, the resulting model closely tracked the complex, non-monotonic behavior that exists in practice. This is ultimately the test of validity for any simulator model.[17]

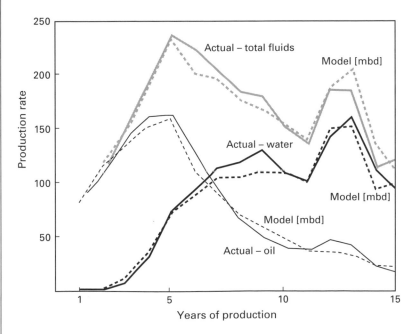

Figure 5.2
Comparison of simulator and detailed models.
Source: Lin, 2009.

decrease the utilization of operating theaters and expensive specialist personnel and equipment.

It takes care and effort to develop insightful top-down models. The work requires two types of investigation. First, we need to document chains of physical interactions to answer the question: How do changes in one area affect operations in others? At the same time, it is necessary to develop an understanding of how the humans in the system—doctors, nurses, administrators, and patients—react to the physical changes. For example, cancellations of elective procedures can give a hospital a bad reputation for elective care and depress demand. Conversely, more patients might choose to use the hospital once they learn that its services have improved. In general, good top-down models require an integrated perspective on both the mechanical and behavioral aspects of the system.

Top-down models are particularly useful in situations where a system involves feedback over time between its components. As the hospital example suggests, a change in one area of the system frequently has complicated knock-on impacts on quite different areas, often with some delay. For example, service improvements are unlikely to increase demand immediately; it takes time for the word to get around and for people to react to the new situation. Such delayed responses are especially important for suppliers of goods and services. A large manufacturer, for example, may have a base of loyal customers who are locked in by habit or contract; if the company fails to maintain competitiveness, a loss in sales will take time to manifest itself fully. Designers need to understand such contexts because they affect the markets for their products and profitability.

In general, changes to a complex system are rarely limited to immediate effects. Normally, and especially when humans are an integral part of the system, developments in one area change conditions for participants in other areas, which in turn ultimately affects their behavior. It is important to understand these ripple effects. We cannot plan intelligently without thinking about the sequence of potential consequences of our actions.

Systems dynamics has proved to be a good way to present top-down models with feedback and delayed responses. This is a modeling procedure based on 50 years of development and supported by a range of standard software.[9] Its design makes it easy to describe the interactions between different elements of a system, both mathematically for the computer and visually, so that planners and designers can appreciate the way that changes in one part of the system affect other parts (see box 5.8).

Use of Screening Models to Identify Candidate Flexibilities

We can use screening models to identify candidate flexibilities in three general ways: conceptual, optimization, and patterned search.[10] The conceptual approach relies on the simplicity of the model. It exploits this feature to enable analysts and decision makers to think through the issues and agree on possible designs for in-depth analysis. Optimization search is feasible in special cases where the model structure makes it possible to apply mathematical tools to home in on likely best designs. Patterned search uses simulation to explore many alternatives systematically, can be widely applied, and is the most general technique of the three. The best approach depends on the problem and situation, but they

Box 5.8
Top-down systems dynamics model: Electric power in Kenya

The development of electric power in Kenya has been problematic. The national power company has found it difficult to finance enough production capacity and a distribution grid that will provide reliable power. This encourages industry to build duplicate backup facilities to ensure production. Customers meanwhile invest in solar energy and diesel generators, thus reducing the potential markets for the national company and discouraging it from connecting villages to the national grid. The fundamental question is: How should Kenya best develop its electric power system?

Steel developed a top-down model of the Kenya power system[18] based on 6 months of field study, assembling technical information about the system and conducting interviews to understand the behavior of power producers, industrial users, and individual customers. Figure 5.3 shows a portion of her model. It maps out the factors that describe the way industry in Kenya could decide on their strategy for obtaining power from either the national grid or their own sources.

Steel used the complete model to show the effect of different development strategies on the future market for electric power in Kenya. Some scenarios might lead to an integrated national grid, whereas others would favor a decentralized pattern of local production and use. This kind of information helps planners and developers choose to develop the kinds of projects that will best add value. Figure 5.4 shows one of the possible scenarios.

Box 5.8
(continued)

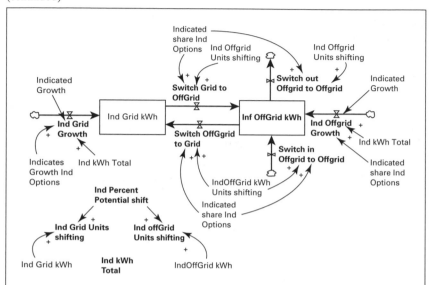

Figure 5.3
Systems dynamics model of industrial decisions concerning electric power in Kenya.
Source: adapted from Steel, 2008, p.199.

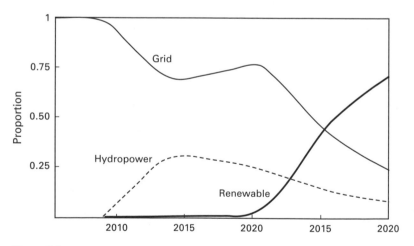

Figure 5.4
Possible future market shares in Kenya among grid, hydropower, and renewable
electrical energy. This scenario might develop if unreliability in the grid encourages
the development of distributed local renewable electricity.
Source: adapted from Steel, 2008, p.169.

can be combined. For example, the conceptual approach can guide a patterned search.

Conceptual Approach

The conceptual approach is particularly useful in getting planners and designers to "think outside the box." An overall perspective shows the interactions between the different functions of the systems. It brings these interactions into focus for the specialists and experts in each particular function, and it allows them to appreciate how their performance depends on that of others. It opens up opportunities for exploration of new design opportunities and candidate flexibilities.

The example in box 5.9 illustrates how the conceptual approach works. The links between the distinct functions of a system may be common knowledge, yet the experts in these functions often ignore them in practice because they are outside the experts' immediate areas of expertise or responsibility. In such situations, screening models play an important part in spotlighting the importance of functional interdependencies.

Box 5.9
Conceptual approach to using screening models: Oil field development

Screening models that integrate several aspects of a complex design can help professionals see beyond their immediate responsibilities and concerns. We observed this in the planning for the development of an oil field.

The development team focused on platform design. However, a larger view of the situation indicated that the performance of the field, and thus of the platform, depended significantly on the way the undersea wellheads connected with each other. These tiebacks between wells allow the managers of the oil field to distribute the flows from the wells according to their quantity and viscosity and improve the performance of the oil field and, by extension, the value of the platforms. In this case, incorporating the consideration of tiebacks in the design helped increase the expected value by 78 percent![19]

The platform designers knew that management of the tiebacks had a strong effect on the platform's performance. However, until the MIT team had developed the screening models, and made the influence specific, the design process for the platform kept to its "box" and did not consider the effect of these underwater elements on the platform's performance. The screening model enabled the design team to "think outside the box."

Optimization Approach

Optimization works best when applied to systems where there is a single coherent model to which some optimization method can be applied. Such models exist in many fields, particularly those that can be represented by networks of flows. A supply-chain network connecting factories, warehouses, and end users is one example. Others involve the distribution of electric power or the development of a river basin.

The situations for which the optimization approach works well contrast with other common, even prevalent, fields in which the system model is an assembly of very different models. The oil-and-gas model for the evaluation of oil field development is one example. This combines distinct models for the flow of oil and gas through porous media, for the operation of oil platforms and related mechanical processes, and for the economic evaluation of the value of the entire system. Similarly, the models for communication satellites combine distinct technological elements for the use of information bandwidth, the deployment of satellites, the estimation of future demands for service, and the overall value of the network.

It is important to understand that optimization on the basis of a single projection of the future will lead to a solution that is optimal for that one future but typically bad for other distinct possible futures that do not play a role in the optimization process. In this sense, naïve optimization amplifies the problems with single number projections. Nevertheless, optimization can be useful to identify candidate flexibilities if applied in the following way:

1. Optimize the design for one set of contextual conditions (such as estimates of future demand and prices) normally chosen in the midrange of the range of values expected.

2. Repeat the optimization process for several levels of the major contextual conditions to observe how the optimal solution changes.

3. Observe which elements of the optimal design change when the contextual conditions change—these are the ones where we will find the flexibility to adjust the design. Conversely, design elements that are insensitive to changes in the contextual conditions do not need to be flexible.

The planning for the development of a river basin in China (see box 5.10) illustrates this approach.

Box 5.10
Optimal approach to using screening models: River basin in China

This case involved great uncertainty in the long-term price for hydroelectric power. China was shifting to market pricing from a regime in which government agencies set prices with little consideration for the cost of production or the willingness to pay. Correspondingly, we could expect the schedule of prices to shift to some new norms. Future demand for power was then doubly uncertain because uncertainty about prices increased the difficulty of estimating long-term regional growth.

The analysis team used the screening model to identify what part of the design should be flexible. They specifically explored the effect of uncertainty in the price of electricity. Variability in this important factor not only influences the value of the project but also, as is generally the case, significantly changes the optimal design. In this case, it affects the optimal size of the development at site 3. Conversely, the designs for dams at sites 1 and 2 were not sensitive to the price of electricity, as table 5.1 shows. The analysts thus successfully screened the plans to identify the design element that should be flexible, that is, site 3.

Table 5.2 completes the story. The screening process focused attention on keeping the design for site 3 flexible until some future time when uncertainties about the prevailing price of electricity were resolved. In addition, the analysis demonstrated that the value of the design was time-sensitive so that the developers should be flexible about when they start elements of the project.

Table 5.1
Optimal system designs for different prices of electric power product

Price of electric power Renminbi/ KWH	Value of project Renminbi/ millions	Optimal characteristics of dams					
		Site 1		Site 2		Site 3	
		Power MW	Volume 10^9 M^3	Power MW	Volume 10^9 M^3	Power MW	Volume 10^9 M^3
0.10	0	0	0	0	0	0	0
0.13	367	3,600	9,600	1,700	25	0	0
0.16	796	Same	Same	Same	Same	0	0
0.19	853	Same	Same	Same	Same	1,564	6,593
0.22	1,607	Same	Same	Same	Same	1,723	9,593
0.25	2,196	Same	Same	Same	Same	1,946	12,242
0.28	2,796	Same	Same	Same	Same	1,966	12,500
0.31	3,396	Same	Same	Same	Same	1,966	12,500

Source: Adapted from Wang (2005, p. 188).

Box 5.10
(continued)

Table 5.2
Sources of flexibility value

| | Source of flexibility value | |
Site	Timing	Design
1	Yes	NO
2	Yes	NO
3	Yes	Yes

Source: Adapted from Wang (2005, p. 189).

Patterned Search

A patterned search is similar to an optimization search insofar as it systematically tries out different types of design. The difference is that the patterned search is not directed by a set of procedures that lead it to an optimum solution; it does not benefit from a mathematical process to determine which designs might be optimal. The designs explored by a patterned search depend on guidance developed from conceptual models with which the design team are familiar or from similar cases for comparable projects .

A patterned search differs from a conceptual search in that it hunts mechanically for desirable flexibilities. By contrast, in a conceptual search, flexibilities become evident as design professionals think outside their professional domains and consider the whole, as the examples in box 5.9 suggest.

In the absence of any specific guidance, the default is to consider standard alternatives for providing flexibility. This straightforward process tries out flexible designs that have proved to be effective in other situations. These include:

• *Phased design* If system operators expand capacity in small units, they can limit capacity if it turns out not to be needed and can time expansion according to the growth in demand. This approach has the advantage of deferring costs, if not eliminating them entirely, and thus of increasing present values. Building modules also reduces costs as developers learn how to produce units more effectively. However, building small may forgo economies of scale. In appendix C, we discuss the interactions of these effects in detail.

• *Modular design*   Design the system for "plug-and-play," that is, with the capability to accommodate the addition of new features through simple connections. For example, USB ports allow computer users to add all kinds of capabilities to their existing systems.

• *Design for expansion*   In designing for expansion, systems are created with the built-in capacity to expand in size. Prime examples of this are bridges originally built with the strength to carry an eventual second deck should the demand arise, such as the George Washington Bridge across the Hudson River from New York City and the Ponte 25 de Abril across the Tagus in Lisbon.

• *Platform design*   A "platform" is a basis on which the system managers can create many different designs, already conceived or yet to be developed. This approach is well known in major consumer goods manufacturing. For example, automobile assemblers routinely put several different styles of chassis on top of a common base and wheel train.[11] Likewise, makers of power tools often have a standard core, consisting of a grip and motor, to which they attach different kinds of business ends.

• *Shell design*   This involves designers creating the capacity for some future use without immediately dedicating it to any particular use. In a number of cases, hospitals have built extra space for which they have no immediate use, but which they can later fit out for offices, wards, or even operating rooms. They recognize a generic future need, but are reluctant to commit to expensive completion until future requirements become evident.

Screening models often deliver excellent results in terms of increasing the expected value of the system. This is because they help identify flexible designs that enable system managers to adapt efficiently to what they may eventually need while avoiding the losses associated with investments that turn out to be superfluous. Flexible designs both increase the expectation of gains and reduce possible losses, tending to produce net gains in expected value.

Example Application: Configuration of Automobile Factories

Automobile manufacturers need to decide which plants to equip for the production of which models. Furthermore, they have to make these choices 3 or so years in advance, well before they can know how many cars of what type they will sell during the life of their production lines.

The standard way to deal with this dilemma is to work with the best estimates of future sales of each type of car, and it is fundamentally flawed. Analysts formulate an optimization problem to determine an "optimal" allocation of capacity to produce cars of different types to multiple plants over several years. Conceptually, this process is like designing a parking garage based on a best estimate of future demand. As the parking garage example described in chapter 3 and box 3.1, the standard process is subject to the flaw of averages. It is almost certain to develop a design that is inferior to what we could achieve if the analysis recognized uncertainty and enabled a flexible design that could respond effectively to actual future sales of different types of cars.

A good flexible solution for the allocation of capacity to plants should provide interesting opportunities for reducing costs and increasing profitable sales. Working with a major automobile manufacturer, a team from the MIT Materials Systems Laboratory set out to identify good flexible designs. Their trial attempts to optimize the allocation, considering the possible stochastic variability of demand over time, demonstrated that this approach entailed excessive computation time. The team thus developed a screening model to identify desirable flexible options for the production of different cars in plants of various sizes.

The team considered three distinct major issues concerning the production process: the allocation of car types to plants, the capacity to produce each type of car in the plant, and the possibility of stretching this production by using overtime in the actual manufacturing process. A different form of analysis is appropriate for each level, as figure 5.5 indicates:

• The consideration of different allocations of capacity is at the highest level. In this case, the analysis explored the range of possibilities using the "one-factor-at-a-time" experimental design;[12]

• Next, the analysis used a screening model to explore the value that could be achieved with different decisions about capacity for each plant over time; and

• Finally, for any allocation of production capacity to plants, the process optimized the use of overtime for the possible range of demand levels.

The overall analysis incorporated detailed submodels representing the cost of production for each of the different processes in the plant. The MIT Materials Systems Laboratory had worked these out in detail through years of collaborative research with major car manufacturers.

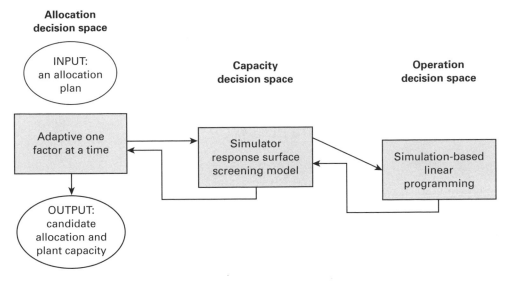

Figure 5.5
Structure of the analysis to identify the interesting flexible designs of the allocation of capacity to automobile plants.
Source: adapted from Yang, 2009.

Yang, a researcher in the Laboratory, developed a simulator screening model to explore all the combinations efficiently.[13] Figure 5.6 illustrates the procedure she used. For visual clarity, it represents only two design choices: the size of the capacity in each of two plants. However, the actual process considered many more factories. To develop the simulator, she selected combinations of capacity allocations—represented by dots on the bottom plane of figure 5.6—and calculated the ENPV that each combination would deliver, indicated by the corresponding points above plane. She then used statistical analysis to fit a functional form to the results; this is the parabolic cone in figure 5.6. She then used this function as a basis for optimizing the combination of operational procedures and capacity allocations. By using this simulator screening model, Yang was able to develop a reasonable estimate of the value of alternative designs for the entire manufacturing process. By applying this analysis to possible allocations of car types to plants, as figure 5.5 indicates, she was able to identify a short list of flexible designs that could be investigated in complete detail.

The use of the screening model delivered impressive results. As table 5.3 indicates, the process identified flexible designs that delivered better economic performance. ENPV rose by almost 50 percent, leading to a

Response surface

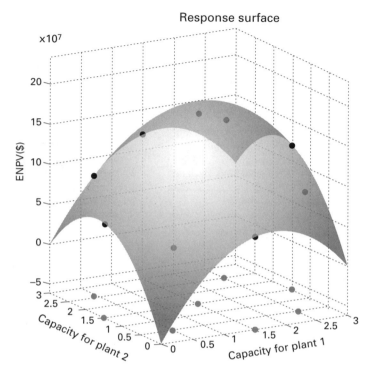

Figure 5.6
Response surface simulator screening model used by Yang (2009).

Table 5.3
Screening model delivers significant improvements in performance

Design choice	Investment, $, millions	Net present value, $ millions			Capacity utilization, percent	Return on investment, percent
		Expected	P₅	P₉₅		
Plan DA1	340	68.4	28	98	98	20.1
Plan DA2	**324**	64.8	27	92	**101**	20.0
Plan DA3	359	91.3	60	**115**	95	25.4
Best Flexible	346	**93.8**	**64**	114	**102**	**27.1**

Best performance in each category in highlighted in bold.
Source: Adapted from Yang (2009, p. 159).

35 percent jump in the return on investment (from around 20 to 27 percent). Meanwhile the possibility of poor returns improved significantly; the P_5—the level under which only 5 percent of the possible outcomes could happen—rose 125 percent from \$27 million to \$64 million. Moreover, the P_{95} for the best flexible design rose about 20 percent, from \$98 million to about \$114 million. The flexible plan achieved these great improvements by adapting expensive capacity closely to actual sales as they developed, as indicated by its average capacity utilization of 102 percent (achieved by using overtime as needed). On balance, it is clear that the use of the screening model successfully identified desirable flexible designs.

Take Away

The recommended approach to identifying the most valuable kinds of flexibility for a system is to use screening models. These are simple representations of system performance that require little time to run. They contrast with the complex models used for detailed design, which may take days to run, and constrain designers to examining only a few alternatives under limited conditions. Screening models enable designers to analyze the performance of alternative developments under all kinds of conditions. Screening models provide the way to determine which flexible designs may offer the greatest value.

Screening models come in different forms. They may be bottom-up, simplified versions of the complex, detailed models used for design. They may be simulator models that statistically mimic the complex models. They may be top-down representations of major relationships between the various parts of the system. The most appropriate approach depends on the situation and what is available.

Analysts will identify good candidate flexibilities by exercising screening models. Sometimes the analysis will be conceptual. Analysts, system managers, and other stakeholders in the system will use the screening model of the system as the basis for thinking outside their professional boxes to see how flexibilities in the overall system could be beneficial. Most often, designers will use screening models to test possible flexibilities over many combinations of possible uncertainties.

The overall result is to define good candidate flexibilities, that is, a short list of possibilities. Because these are developed using screening models, which are deliberately simple and incomplete, they will need validation using the standard processes for detailed design.

Asking the right question is half the answer.
—Old saying

In this chapter, we provide a process for choosing the preferable solution in the context of uncertainty, expanding on standard procedures for evaluating projects. Although we can apply the proposed methodology to a wide range of system performance metrics, economic value, estimated by discounted cash flows (DCF) and net present values (NPV), are at the core of this chapter. Appendix B briefly reviews DCF and NPV. These metrics, however, are neither accurate nor sufficient once we recognize that future contexts and outcomes are uncertain. Thus, the proper evaluation of projects with uncertain outcomes extends the evaluation procedure beyond the simple formulas that assume we can know the future.

The process for evaluating and choosing designs in the context of uncertainty extends standard evaluation procedures in three ways. It recognizes that:

• The evaluation must consider the range of scenarios, not only the most likely futures. It thus avoids the flaw of averages (see chapters 1 and 2 and appendix A) and obtains a more accurate assessment of value.

• The value of a project is a distribution of possibilities. We must therefore think in terms of *expected* net present values (ENPV) over a distribution, rather than a fixed, single NPV result.

• The complete evaluation needs to consider several factors. This is because two very different distributions of value may have the same expected value. A fair comparison of projects must then characterize the underlying distribution of economic value in some way, for example, by describing upper and lower extremes for each project. A good evaluation should be multidimensional.

The process for evaluating and choosing the preferable project or design incorporates these considerations.

Note the stress on choosing "preferable" rather than "best" solutions. It is important to keep in mind that it is unrealistic and unhelpful to think of best designs once we understand that the goodness of a project has several distinct attributes. This is because different decision makers define "good" and "better" according to their own preferences. For example, one person may prefer a project with higher rewards even though it is more risky than the alternative, whereas another may need to avoid risky endeavors. Furthermore, when a project concerns several stakeholders or decision makers, the decision they agree to (and so "prefer" to others in operational terms) depends on their relative power and on the procedures by which they arrive at their agreements or compromises. In general, whenever we choose between projects with multidimensional values, we should think in terms of preferred choices and recognize that all individual stakeholders have their own.

There is a three-step process for evaluating and choosing designs:

• *Evaluation of individual designs*   In the context of uncertainty, this analysis has two special features. First, it involves consideration of the various distributions or scenarios of possible futures. Second, it requires the analyst to define the circumstances in which system operators will use the flexibility in a design.

• *Multidimensional comparison of designs*   This analysis calls for clear presentation of the different values of each project. The idea is to help decision makers focus on the issues that are most important to them and to help them determine their choices.

• *Validation by sensitivity analysis*   Because estimates of future uncertainties are imprecise, responsible analysis will explore the consequences of different assumptions about the future. As far as possible, we need to develop confidence that choices rest reliably on the analysis.

This chapter develops these steps sequentially. We end with a case study giving an important practical example of the design and development of a major oil field.

Evaluation of Individual Projects

Distribution of Outcome Values

It is essential to analyze each candidate design over the range of important uncertainties. If we do not do so, if we consider alternatives only from the perspective of the expected or most likely set of future circumstances, our calculations of system value will almost certainly be seriously wrong. Failure to examine the performance of designs under the range of circumstances is to fall into the trap of the flaw of averages. If we fixate on the most likely circumstances, our analysis will miss the effect of uncertainties, and we are likely to settle on inferior choices.

The analysis considering distributions of events will give a distribution of possible values. Combinations of most likely events should give more likely outcomes. Combinations of favorable circumstances should lead to particularly good results. Likewise, combinations of unfavorable events should lead to bad results. The results of combinations of favorable and unfavorable circumstances will lead to results somewhere in between, in ways that are difficult to anticipate intuitively. We must do a careful analysis to obtain a useful assessment of the distribution of value for any project.

Monte Carlo simulation is the recommended process for analyzing the range of possible outcomes for design alternatives. It is a standard process that consists of:

• Estimating the range of possible realizations of the uncertain variables that characterize the environment of the system, as well as the relative likelihoods of these realizations (see chapter 4); and

• Evaluating the performance of a proposed design many times for the range of possible uncertainties according to their estimated probability of occurrence.

Importantly, the analysis develops a distribution of possible performance for the trial design. Box 6.1 gives an example.

Mathematically speaking, Monte Carlo simulation calculates the performance of each design alternative considering the joint distribution of uncertainties. It does this in two phases. First, it samples the distributions of possible circumstances (such as future demand, prices, etc.). This gives one possible result. Second, it repeats the sampling process a great many times (e.g., 1,000), giving each possible future circumstance its appropriate chance of being sampled. It thus creates the distribution of the performance of the design that is consistent with the joint distribution of possible circumstances.

Box 6.1
Simple simulation: Demand for parking spaces

Consider the financial evaluation of the parking garage mentioned in chapter 3. Its future annual demand is uncertain. The standard model is a spreadsheet of revenues, costs, net revenues, and present values. The basic evaluation process inserts the demand for each year into the spreadsheet, which generates the revenues, operating costs, net cash flows, and an NPV.

The Monte Carlo simulation uses the basic process as many times as desired, perhaps a thousand. Each time it generates and applies a set of possible demands for parking, for each year over the life of the project as generated by the distribution of possibilities. For example, the analysis might assume that we sample year-on-year changes in traffic from a bell-shaped distribution (also known as the Normal distribution). The simulation would then generate a sequence of possible future scenarios, such as those shown in figure 6.1.

From each set of yearly traffic flows, the simulation generates an NPV. From the distribution of possible traffic flows, it thus generates a distribution of NPVs, as figure 6.2 shows. Note that this distribution is not bell-shaped like the input. As usually happens, the physical features of the project distort the results. In the case of a garage with limited capacity, the higher levels of traffic do not result in correspondingly higher NPV, beyond that associated with a full garage, because capacity is limited.

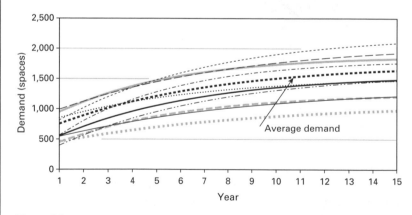

Figure 6.1
Possible future demands for parking spaces over the 15-year life of the garage, generated by the Monte Carlo analysis.

Box 6.1
(continued)

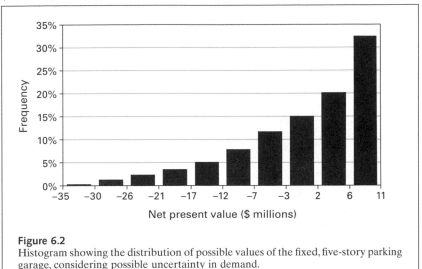

Figure 6.2
Histogram showing the distribution of possible values of the fixed, five-story parking garage, considering possible uncertainty in demand.

The simulation keeps a record of everything it does so that sensitivity analyses can examine the results in detail, as we discuss later. Appendix D describes Monte Carlo simulation in detail. The software is widely available, inexpensive, and easy to use.[1]

Use of Screening Models

Monte Carlo simulations routinely involve 1,000 analyses of a design and often many more.[2] We need numerous values to obtain a good representation of the distribution of the values associated with a design. To carry out the calculations in reasonable time, we want to work with models of the system that need as little time as possible to run a single simulation. Monte Carlo simulations predominantly use screening models, the simpler, mid-fidelity representations of the system described in chapter 5.

Monte Carlo simulations based on screening models are appropriate in the initial phase of the design when we are developing a short list of the best possible forms of flexibility to include in the architecture of the system. In this phase, when we are identifying overall concepts for system design, we do not have to worry especially about detailed accuracy because we will later carry out high-fidelity analyses of the possibilities short-listed.

However, we are unlikely to be able to use the standard Monte Carlo analysis when we come to the detailed design because the models needed for that take too long to allow for thousands of simulations. At this stage, we will have to be satisfied with far fewer tests of the design, and we will have a cruder description of the performance distribution of the alternative design choices. Nevertheless, this does not alter the fundamental reality that we need to compare distributions of values for each design.

Rules for Exercising Flexibility

The evaluation of a flexible design differs from that of a fixed design. Because the fixed design is passive and simply reacts to future circumstances, the analysis of a fixed design is simple. By definition and construction, a fixed design does not change over its lifetime.

In contrast, we expect a flexible design to change over its lifetime. Indeed, the whole point of a flexible design is to enable management to respond proactively to circumstances as they evolve. The evaluation of a flexible design is thus more complex than that of a fixed design; the valuation model has to account for the possibility of design changes and their effects.

The issue in evaluating a flexible design is that we generally cannot know in advance when management will change the design—because these decisions will depend on yet unknown future conditions. If the demand for project services increases rapidly, management may quickly exercise the flexibility to expand the system. However, if the demand increases slowly, management may decide to expand slowly, late, or never at all. In short, we can expect management to take advantage of flexibility in a design when circumstances are "right." The Monte Carlo evaluation of a flexible design has to incorporate some procedures to identify when the possible future circumstances are "right" and then change the design so that the evaluation can mimic intelligent system management.

The "rules for exercising flexibility" provide the mechanism for identifying when and how intelligent management would take advantage of flexibility and change the design. These rules monitor each possible future scenario or forecast developed by the Monte Carlo simulation, look for situations that call for design changes, and instruct the automated analysis to adapt the design and system performance in the basic evaluation model. The rules for exercising flexibility should mimic what management would do if they ever had to deal with a situation. Good rules for exercising flexibility are thus essential for the assessment of the value of a flexible design.

A "rule for exercising flexibility" is a small logical element that, as its name implies, determines when (and how) the system should change in response to changing context and opportunities. Using terms from computer programming, the rules for exercising flexibility are "IF, THEN" statements. For example, if management only wants to expand capacity once demand for additional space appears robust, then its rule might be to expand only after observed demand exceeds capacity for at least 2 consecutive years. A suitable rule concerning the possibility of adding more floors to a parking garage would then be:

IF... demand exceeds capacity for 2 consecutive years,

THEN add floors to meet demonstrated demand.

This rule would keep track of the relationship between demand and capacity for 2 years before triggering an expansion, the associated costs of construction and operation, and the benefits of increased performance. Such rules are easy to program into a spreadsheet, as we explain in appendix D on the Monte Carlo simulation.

The simulation process calls on the rules for exercising flexibility each time it generates new values for the uncertain variables, that is, in each period of the simulation. At that point, it consults the rule for exercising flexibility and asks: "Are the circumstances right for exercising the flexibility?" If the answer is "No," the simulation does not change the description of the system (e.g., the spreadsheet). If, however, the answer is "Yes," the rule triggers the change in the system and continues the subsequent analysis with that change in place. In chapter 3, we used the example of a parking garage that could be expanded if demand were high. An appropriate rule for exercising flexibility would, for example, check whether the simulated demand in the previous 2 years exceeded capacity and add extra floors if that were the case.

In line with what could happen in reality, any rule for exercising flexibility leads to a wide range of development paths. Taking the parking garage as an example, in some scenarios, the traffic does not grow, and the garage does not expand. In others, demand grows rapidly, and the simulation accordingly adds floors early in the life of the project. In general, the rules lead to a distribution of developments, each with their own NPV.

We could of course apply different rules for exercising flexibility at different times over the life of a project. For example, management might be ready to expand a project during the early stages of the project but not want to spend money on changing the system as it nears the end of its useful life. Analysts can easily program different rules into their simulations. The essential thing is to anticipate management choices as reasonably as possible.

Target Curves

A "target curve" is a convenient way to present the distribution of possible values associated with any design. It shows the probability that realized performance will be lower than any specified level or target. It derives directly from the results of a simulation. We achieve this by sorting the generated performance values in ascending order. Thus, figure 6.3 is the target curve associated with the distribution in figure 6.2.

It is standard to indicate the average value of the distribution as a vertical reference line on the graph. Figure 6.3 thus shows that the ENPV of the project is $0 for this case.[3] Note that, as in figure 6.3, the expected value does not necessarily occur in the middle of the distribution. Because the distribution of outcomes is generally asymmetric, we can anticipate that the ENPV will be at some other point than the median, the target with a 50 percent probability.

A target curve presents a lot of information in a compact form. Referring to figure 6.3, it shows:

• The probability of breaking even (NPV = 0) or doing better on the project, which is the complement of the probability of making a loss (in this case, the probability of at least breaking even is 60 percent);

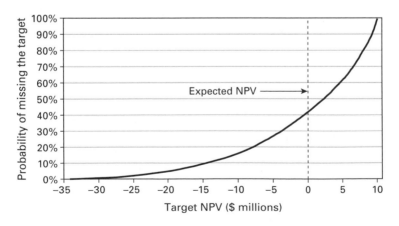

Figure 6.3
Target or cumulative percentile curve for a fixed, five-story parking garage.

Table 6.1
Measures of distribution of outcomes of a project design

Metric	Present value
ENPV	$0 million
Probability of breaking even	60 percent
10 percent value at risk or P_{10}	$15 million loss
90 percent value at risk or P_{90}	$9 million gain
Minimum result	$34 million loss
Maximum result	$10 million gain
Range of results	$44 million span
Difference between median and ENPV	$2 million

• The range of results, reflecting the dispersion in outcomes (a span of $44 million);

• The risk of the downside of any specified level (sometimes referred to as the "value at risk"). For example, there is a 10 percent chance of losing more than $15 million and a 90 percent chance that the results will be less than a gain of $9 million. These 10 percent and 90 percent results are also called the P_{10} and P_{90} values. These values are often preferred as measures of the range because they are statistically more stable than the absolute minimum and maximum values generated by a simulation; and

• The difference between the median value at a cumulative probability of 50 percent (a gain of about $2 million) and the average value (0) caused by the asymmetry in distribution, in this case skewed toward great downside losses.[4]

Depending on the context, some of these distribution characteristics may be among the set of dimensions that decision makers will want to consider when choosing between designs.

Table 6.1 represents a summary of the target curve. A table may be preferable for audiences or managers who are not comfortable with graphs. For the graphically inclined, however, the target curve is a useful and an effective way of comparing different aspects of the values of alternative designs.

Using Target Curves as a Guide to Design
Flexibility in the design generally enhances performance in two complementary ways. Flexibility can:

1. Reduce downside consequences, and

2. Increase upside opportunities.

Target curves provide a visible guide to what flexible designs can do to improve the performance of initial designs, as figure 6.4 and box 6.2 indicate.

Actions to reduce downside consequences increase the expected value of a project by minimizing the downside tail of poor results. They may also increase the minimum result. They act like insurance or "put" options in financial terms.[5] A typical way to reduce the downside risk is to invest in a relatively small project at the beginning and defer expenses until

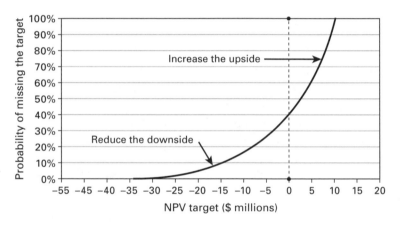

Figure 6.4
Graphical view of desirable ways to improve design by using flexible designs to shift the target curve to the right.

Box 6.2
KEY PARADIGM

> **Improving a system in an uncertain environment means moving its target curve to the right, making the achievement of targets more likely.**
>
> **Designs often have a target curve that we cannot move entirely to the right.**
>
> **Decision makers must make a trade-off between moving parts of the curve to the right, making these targets more achievable, and accepting a leftward shift of other parts of the curve, making the corresponding targets less achievable.**

future development has validated the need for them. For example, with reference to the garage case, it may be better to build only a four-story garage rather than the six-story garage that might be best if we had to choose a fixed design. Obviously, when we limit our investments, we can limit our losses in a project.

Likewise, actions to increase the upside also increase the expected value of a project. They may also increase the extreme upper results, such as the P_{90}. Financially, such efforts act like "call" options. A standard way to increase upside opportunities is to build into a project features that will enable it to morph into a state that meets future demands or requirements. These can be actions that allow for future expansion, such as the early investment in extra strength for the Ponte 25 de Abril across the Tagus, in Portugal, which allowed for later double decking of the bridge when needed. Alternatively, flexible designs may enable a shift from one mode of operation to another, such as when power plants install the capability to burn either natural gas or oil and can thus take advantage of the changes in the relative prices of these fuels.[6]

Comparing Target Curves
Target curves provide a valuable way of revealing the differences between the consequences of different designs. However, they are not easy to compare because they have many features and shapes. They represent a maximum, a minimum, and an average. They may rise sharply and then level out—or the contrary. In short, we need to interpret them carefully.

A target curve for design A may be completely to the right of the target curve for design B. In this case, A represents an improvement over B, insofar as A reduces the chance of missing any desired performance target. Figure 6.5 shows an example of this for alternative designs for an oil platform complex. Development strategies 5 and 8 are each unambiguously to the right of strategy 3. Designs whose target curves are completely to the right of another have a lower probability of failing to meet any target and thus offer a higher expected value.

Note that a design with a target curve completely to the right of another does not necessarily represent a better choice. This is because a target curve does not represent all the important attributes of a design. For example, the design whose target curve is to the right, and thus offers better performance, may also require a greater initial capital investment (Capex). Decision makers might prefer the design that costs less even though it also performs less well. In addition, target curves represent the

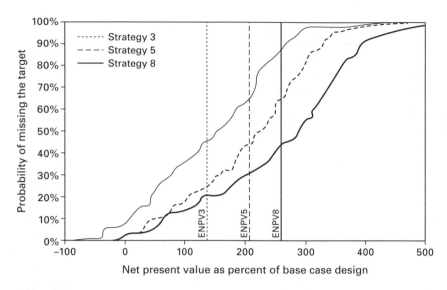

Figure 6.5
Dominant target curves. Strategies 5 and 8 for the design of an oil platform complex dominate strategy 3 (horizontal normalized to preserve commercial confidentiality). *Source:* Lin, 2009.

risk profiles of the economic performance of designs but not other factors—such as environmental performance, which might be crucial to the final selection of a project.

Note also that a design whose target curve is completely to the right of another may not perform better under all circumstances. It may perform exceptionally well in some limited cases, perhaps precisely in those circumstances in which the alternative does not perform well.[7] Target curves represent risk profiles over all cases, not a head-to-head comparison of performance in specific circumstances.

In general, the target curves for alternative designs do not dominate each other. Figure 6.6 illustrates the phenomenon, showing the target curves for designs with different numbers of stories for the parking garage. Comparing the curve for the seven-level garage with that for the four-level structure, we can see that they cross at the level of 50 percent. Until that point, the curve for the four-level garage is below and to the right of that for the seven-level, meaning that the smaller structure has less chance of delivering low performance values. This is of course what we would expect: Having spent less to build the smaller structure, the owners have less to lose. Conversely, above the crossover point, the curve

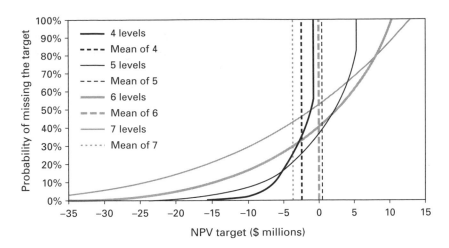

Figure 6.6
Overlapping target curves associated with fixed (inflexible) designs of parking garage with different number of levels.

for the seven-level garage is to the right of the curve for the four-level structure, which becomes vertical. This means that the limited capacity of the smaller garage limits its maximum value, whereas the larger structure can deliver much more value if the conditions are right. This kind of crossover effect is common.

Choosing Between Target Curves

Sometimes the choice between designs seems obvious. Consider the comparison between the target curves for the four- and five-level designs in figure 6.6. The curve for the five-level design is mostly to the right of that for the four-level structure. The five-level design offers significant improvement on the four-level design in terms of its ability to meet targets. It has a 65 percent chance of breaking even or doing better, whereas the four-level design has no chance. Overall, the ENPV of the larger fixed garage is about 0 compared with a loss of $2 million for the smaller project. So it seems clear to prefer the five-level design over the four-level design.

However, the choice between designs is often not obvious. Although the five-level design dominates the four-level one, the choice between five or six levels is not as clear-cut. The corresponding curves cross each other in the middle. The six-level design has considerably higher upside, with a 20 percent chance of achieving a value of $8 million, which is not possible for the five-level design. However, this increased upside comes with an increased downside: The six-level design has a 15 percent chance of losing $10 million or more, against about a 6 percent chance for the five-level design. The choice between these two designs depends on how the system's owners or designers feel about these risks. In general, we need to face up to the fact that choices between designs require us to consider conflicting objectives.

Multidimensional Comparison of Projects

Decision makers generally want a project to perform well over several criteria. For example, they might want the system to provide good value as represented by ENPV, to minimize risks (perhaps as defined as the P_{10} value), and to perform well environmentally. The comparison and eventual selection of projects involve many dimensions.

Trade-offs

In choosing between projects with different levels of attainment, stakeholders balance the advantages and disadvantages of one alternative compared with another. For example, they might compare the return or ENPV of an alternative and the risk it might entail. That is, they "trade off" the different dimensions of projects.

The trade-off that decision makers are prepared to make between two dimensions of choice implicitly defines the relative value they accord these dimensions. For example, we can interpret the extra risk they might be willing to incur for an increase in expected value as their price of risk. If there were some way to define the relative value that decision makers ascribe to each dimension of choice, it would be possible to define an overall measure of value for all of them. This approach is not practical, however.[8]

Good practice considers the various criteria for choice individually. It is not practical to give consistent and accurate monetary (or other) values to so many dimensions of choice or otherwise blend them into a single measure of goodness. It is therefore not possible to define an objective function suitable for overall optimization. Consequently, it is not possible to determine preferred design through a purely mathematical procedure that ranks projects unambiguously. In comparing alternatives, decision makers need to confront several criteria and make the trade-offs appropriate to their situation and preferences.

Concept of Multidimensional Choice

In comparing alternatives, it is helpful to divide them into two categories. Dominated alternatives are those that are demonstrably inferior because they do not perform as well as some other alternative on all criteria of choice. We can eliminate these from consideration. Obviously, if we know of a design that is better on all counts (e.g., provides more value, is less risky, and requires less investment), we will prefer it without further discussion.

Our attention should focus on the dominant alternatives. Unlike a dominated alternative, for which there is always another alternative that is better in all criteria, we cannot improve these designs in any single dimension without giving up performance in some other. With reference to figure 6.7, the dominant alternatives are those on the outer edge of the feasible region of alternatives—in this case, on the curve from A to B. This boundary is the so-called "Pareto frontier" and defines the best available designs. Note that, by convention, the axes all plot benefits so that better performance is to the right and upward from the origin. Thus, alternatives to the left and toward the origin of curve A–B, such as point C, are dominated designs. By construction, there are no known designs to the right and above the Pareto frontier. The concept of the Pareto frontier is general. Although figure 6.7 shows it in two dimensions for simplicity, it can have as many dimensions as necessary. As can be seen on the graph, any increase in one benefit for a dominant design (i.e., one

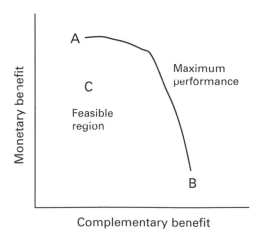

Figure 6.7
General diagram of the performance of designs when all axes represent benefits. In this case, the set of best designs, known as the Pareto frontier, is furthest away from the origin.

on the Pareto frontier) comes at the expense of a decrease in some complementary benefit. The Pareto frontier defines the trade-offs that are available for any dominant design.

The choice between dominant designs depends on the currently prevailing preferences of decision makers and stakeholders. As noted earlier, people's preferences depend on their circumstances. For example, if they are secure, they may be ready to accept more risk, but if they are vulnerable, they may be risk-averse. Whether their choice is A or B (or some other dominant design in between) depends on the relative importance that managers attach to the criteria of choice in their circumstances and to the trade-offs they are willing to make. For example, how much of the monetary benefits of A might they be willing to trade off to obtain greater complementary benefits, such as those represented by B?

As a practical matter, it is generally not possible to determine reliably and in advance how anyone will choose between trade-offs. Managers may not be willing or able to define their preferences until confronted with the choices. Moreover, it may be difficult to define important benefits precisely. For example, a national Energy Minister responsible for the implementation of an electric power system will probably need to balance the economics of national investments against the need to distribute the benefits across the country. A system that provides reliable power only to rich urban areas and disregards poor rural ones might not be sustainable, politically or otherwise. The relative value of such benefits is difficult to define under the best circumstances. It is highly unlikely that anyone can express these values accurately for speculative cases.

The choice of preferred overall design requires some sort of process that allows decision makers to select the solution they jointly prefer from a set of best designs. Box 6.3 gives an example.

Capex Issue

Developers of major systems need to pay special attention to the amount of money they must provide to build the first phase of a project. This amount is the initial capital expenditure (Capex). Although it is included along with all other costs in the calculation of NPV, it differs substantially from the other costs that occur over the life of the system.

The initial Capex for a new system:

• Is a large amount of money—the cost of an automotive factory, launch of a new aircraft or satellite system, development of a major oil field, or

Box 6.3
Development of Rio Colorado in Argentina

The Rio Colorado flows through Argentina from the Andes to the Atlantic. Its development involves structures to promote irrigation in the mountain foothills, control floodwaters, and provide water for industry and coastal populations. Designs that allocated water primarily to the richer industrial regions (designs A) contributed the most to the national economy. Those that distributed the water most evenly (designs B) did not create substantial financial benefits.

If it had been practical to tax the coastal regions to transfer their benefits to the interior regions, designs A might have been acceptable. However, this was not possible. Somehow, the designers had to mediate the claims of the various regions. Their solution, which provided the preferable balance between the objectives of financial benefit and fairness of distribution, was around the "knee" of the Pareto frontier, as figure 6.8 indicates. The managers agreed that this was the best design because, compared with A, it greatly increased the fairness of the distribution of benefits without sacrificing economic benefits excessively.[13]

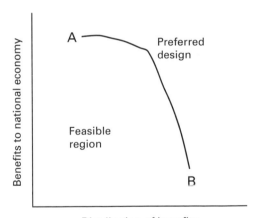

Figure 6.8
The preferred design selected from the set of best designs for the development of the Rio Colorado in Argentina.
Source: adapted from Cohon and Marks, 1973, 1974.

construction of a new highway, bridge, or railroad line may easily exceed $1 billion;

• Has no off-setting revenues—in contrast to operating or expansion costs that are committed during the life of the project and are thus partially offset and paid for by ongoing revenues;

• Cannot be adjusted to the ongoing success of the project—again, in contrast to subsequent operating costs that largely vary with demand, and expansion costs that only occur if sufficient thresholds are met; and

• Is especially risky—because we do not yet know the desirability of the future product or service.

In short, the initial Capex represents an important initial barrier to development—and thus to the choice of a design alternative.

A good analysis should feature the level of initial Capex. This requires special attention. The Monte Carlo evaluation of a system does not give initial Capex a special status; it deals with it as with all other cash flows. It is latent in the analysis, and we need to draw it out and make it explicit.

The obvious way to bring Capex information into the evaluation is to present it, along with the other characteristics of the evaluation, in a comparative display such as table 6.2. This will allow decision makers to consider Capex as they think about the trade-offs involved in making their choice. All else being equal, we can expect that a design with lower initial Capex will be preferred.

Table 6.2
Results, in $ millions, for three possible designs of a hypothetical oilfield

	Three possible designs		
Evaluation metric	Rigid	Less flexible Lower cost	More flexible Higher cost
Expected net present value (ENPV)	822	900	**929**
Standard deviation NPV	285	334	**165**
Low-end result (P_{10})	358	308	**788**
High-end result (P_{90})	**1,286**	1,134	1,148
Expected present value of costs	1,006	**688**	969
Initial capital expenditures (Capex)	625	**419**	589

Best performance in each category is highlighted in bold.
Source: adapted from Hassan and de Neufville (2006).

Case of Uncertainty

In the context of uncertainty, the preferred designs are those that simultaneously increase expected value, particularly compared with fixed designs, and best manage the risks and opportunities in the project. The choice of preferred design embodies several tradeoffs. At a minimum, these are between:

• Initial capital expenditure (Capex) and more speculative later benefits;

• Minimizing downside risks and maximizing upside opportunities; and

• Prospective benefits and possible retrospective regrets about having made the wrong choice.[9]

Different organizations, leaders, and managers will feel differently about these trade-offs, and their views are likely to depend on their circumstances, their competition, and their other major investments.

Illustrative Example

This example illustrates the evaluation over many criteria in the context of uncertainty. The case concerns the development of a major oil field and focuses on comparing a base case against two flexible alternatives:

• A "rigid" design, sized in the standard way, which optimizes the system based on the assumption that the future quantities and price of oil will be as projected;

• A "less flexible, lower cost" design that provides some flexibility and does not deliver the highest ENPV—but whose initial Capex cost is about 30 percent less; and

• A "more flexible, higher cost" alternative that delivers the maximum expected value by exploiting full flexibility to adapt to the actual quantities and price of oil that occur.

Table 6.2 presents the performance of these choices according to important criteria. As it suggests, the choice among the three designs is not obvious: Each design is best in at least one category, as the boldface type indicates.

Both flexible designs have advantages and together seem to dominate the rigid design. However, the case is not definitive. The rigid design, "optimized" to a particular set of circumstances, performs best when these occur. In this case, the rigid design provides the highest possible performance (1,286 in table 6.2) compared with the flexible alternatives that incur the cost of providing flexibility. Under the many other circumstances that may occur, the rigid, "optimized" design is inferior. Being unable to adapt to circumstances, it can neither avoid downside losses nor take advantage of upside opportunities. Thus, the ENPV of the rigid design (822) is lower than those of the flexible designs (900 and 929). Moreover, because the rigid design commits all its capital costs at the beginning, it is by far the most expensive in terms of the expected present value of costs (1,006 compared with 688 and 969). The rigid design neither delays any of the costs (savings in present value) nor avoids them if certain design capacities are not needed (savings in avoided costs). Thus, although the rigid design may be "optimized," it is in fact often an inferior design in the light of the possible scenarios.

The more flexible design might appear most attractive because it provides the highest ENPV (929). However, it is more than 40 percent more expensive than the less flexible design (969 compared with 688) based on the ENPV of total costs that occur over time (e.g., when management exercises the flexibility).

The less flexible design is thus arguably preferable. The more flexible design requires an extra $281 million in expected present value of costs, for an additional gain of only $29 million in terms of overall ENPV. The small gain from the upgrade from the less to the more flexible design does not seem to justify the extra costs, which are about 10 times larger. That is, although the more flexible design is perhaps "best," it may be a "best" that is not worth paying for.

However, the more flexible design in this case is more reliable in terms of delivering results. The standard deviation of its NPV is by far the lowest (165, compared with 285 and 334 for the alternatives). In this sense, its performance is most "robust."[10] Put another way, its minimum result (788) is by far the best—more than twice the value of the alternatives (358 and 308). This result is not surprising because the more flexible design is more able to adjust to the circumstances that may prevail; it is best able to avoid downside results and capable of taking advantage of upside opportunities.

The choice between the more and less flexible alternatives is thus not obvious. Depending on the needs, preferences, and risk perceptions of the decision makers, they might choose one or the other. The analysis helps to quantify the value of flexibility. This enables decision makers to compare this dimension with other criteria and make more informed decisions. The role of the analyst is to bring out the issues and help project leaders make decisions that are preferred from an overall perspective.

Difference Curves

A difference curve provides a useful way to think about the choice between two alternatives. It plots the difference in the target measure of performance between two designs. It helps us think about questions such as: Is higher ENPV worth the extra cost and risks associated with an upgrade from another design? In contrast to target curves, which present overall performance, difference curves highlight the change between two designs. We calculate them by Monte Carlo simulation in the usual way. However, instead of recording the value of one design for each generated future scenario, we now calculate the difference between the values of the two designs for that particular scenario and record these differences for all future scenarios generated.

A difference curve illustrates the relative value of a flexible version of a fixed design. Consider the example of the parking garage in chapter 3. There are two designs for four-level parking garages. The fixed one has no possibility of expansion. The flexible one can extend vertically up to eight levels because its design features, such as stronger columns and footings, will take the added weight of additional levels if they are added. The flexible design costs more than the fixed design for the same number of levels—this increases its possible downside. However, the flexible design can expand its capacity and take advantage of higher demands when they appear. The target curve for the flexible four-level design is thus far to the right of the fixed design in terms of maximum gains. The comparison of the target curves for these two designs in figure 6.9 brings out this advantage of the flexible design. The difference curve in figure 6.10 makes the effect salient.

Figure 6.11 shows the difference between a flexible four-level design and a fixed five-level design. Reading the figure, we can see that the flexible design in this case:

• Delivers a gain in ENPV of $19M,

• Has an 80 percent chance of performing better than the fixed design, as indicated by the break-even point in figure 6.11; and

• Has a 30 percent chance of outperforming the fixed design by $4 million or more, which would seem to outbalance the 20 percent chance of underperforming the fixed design.

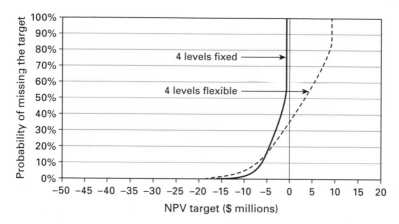

Figure 6.9
Comparison of target curves for fixed and flexible designs of four-level parking garage.

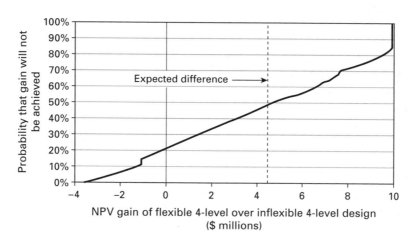

Figure 6.10
Difference curve comparing fixed and flexible designs for four-level parking garage.

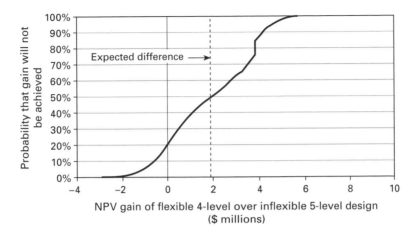

Figure 6.11
Difference curve showing the advantage of the four-level flexible design over the five-level fixed design.

Validation by Sensitivity Analysis

Upside-Downside Curves

Upside-downside curves are a convenient way to bring out the different levels of risk associated with alternative designs. They compare the performance of different designs both on average (i.e., their ENPV) and at selected equal levels of probability, such as P_{10} and P_{90}. They provide a way to compare performance average and spreads.

Figure 6.12 shows the upside-downside curves for the different parking garage designs. The upside line shows the probability (in this case, 10 percent) of achieving a profit as high as or higher than the profit on the line. Conversely, the downside curve shows the probability (in this case, also 10 percent) of having a loss as large as or larger than depicted on the line. The middle line shows the average or ENPV of the design.

In this particular situation, the comparison of performance spread and average values indicates that the five-level design delivers about the same level of average performance (ENPV) as the six-level design but with far less uncertainty—a smaller spread and standard deviation—about the results. Depending on the preferences of the decision makers, this might be a sufficient reason to prefer the five-level fixed design to the six-level fixed design.

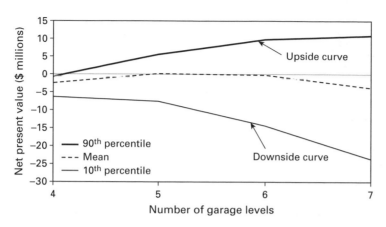

Figure 6.12
Upside-downside curves associated with designs for a different fixed number of levels. The upside curve is for a 10 percent chance of a gain equal or above (the same as 90 percent below) and the downside curve is for a 10 percent chance of results lower than indicated.

Regret Plot

Project leaders may also want to think about possible regrets they might have about their choice between two alternatives—between a fixed and flexible design, for example. They might ask: "If we choose the flexible alternative, what are the chances that we would ever think that we made a bad or wrong choice?"

We can answer this question by creating a regret plot. This device compares the performance of the two alternatives in each of the scenarios generated by the Monte Carlo simulation. Figure 6.13 shows a regret plot comparing the flexible four-level and fixed five-level designs.

Each point in the plot corresponds to one generated scenario in a Monte Carlo simulation. The vertical axis corresponds to the value of the prospective preferred design (the flexible four-level design in this case), whereas the horizontal axis is the value by which this preferred design exceeds that of the alternative. All the situations for which the preferred design performs better than the alternatives are to the right of the zero-line in the plot. All the simulated futures where the preferred design delivers a positive NPV are above the horizontal zero-line.

We can use the regret plot in figure 6.13 to make the following useful observations about the flexible design, which generally:

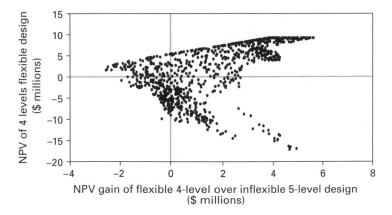

Figure 6.13
Regret plot: case-by-case comparison of flexible four-level design and fixed five-level design for 1,000 demand scenarios.

• *Outperforms the fixed design* Most of the points in the plot are to the right of the vertical split (the Monte Carlo simulation can give you the exact number if desired); and

• *Often delivers positive value even when the fixed alternative would have provided more* These are the points above the horizontal split and to the left of the vertical line. In these cases, the flexible design is not a bad choice, although the fixed alternative would have been better.

Later Capital Expenditures
System managers may want to anticipate when they might find it desirable to commit to second and subsequent phases. Further capital is normally required to take advantage of flexibility, and it is useful to plan for this eventuality. Knowing when they might want to take advantage of flexibility in design helps managers' financial planning.

It may also help them choose their initial design. If the analysis indicates that it is very likely that they will need to implement some flexibility soon, they may want to incorporate this feature in the initial development to avoid any disruption associated with implementing the flexibility. For example, adding more levels to a parking garage is likely to require the owners to close it for some time and, at the very least, work in the midst of operations associated with the structure, such as a shopping mall or office blocks.

Each Monte Carlo simulation has a history of when its decision rules called for the use of flexibility. Keeping track of this history over all the simulations provides distributions of when and how to use flexibility. For example, figure 6.14 indicates that the probability of wanting to expand the four-level flexible parking garage in the first 2 years is more than 40 percent. This information might push decision makers to build a larger facility (such as a five-level flexible garage) from the start and thus avoid disruption so soon after opening.

Monte Carlo simulation can likewise provide information about how much flexibility to use and provides a means of validating the use of flexibility in the design. Thus, figure 6.15 shows the distribution of the final height of the garage. In this case, it shows that the probability of ending up with an eight-level structure is more than 70 percent. However, the analysis also shows that, from the prospective of making a choice at the beginning, a garage with seven or more levels is a losing proposition, in the sense that its expected NPV is negative at the start, as figure 6.12 shows. Putting these observations together means that:

• It would be wrong to build an eight-level or bigger facility at the beginning;

• Yet it is highly probable that it would be advantageous to end up with such a facility; and

• The concept of design flexibility, that is, the ability to move from a good starting point to a different end, is validated.

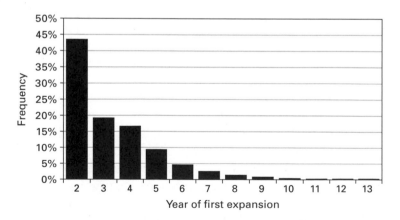

Figure 6.14
Year of first use of flexibility to add levels to parking garage.

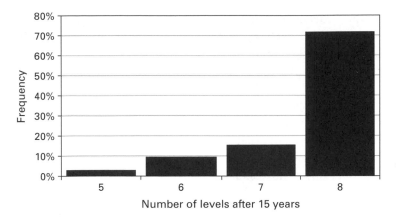

Figure 6.15
Amount of flexibility used over project life of parking garage.

Sensitivity Analysis

Assessments of uncertainty are also uncertain. Our evaluations can thus benefit from an analysis of the sensitivity of the results to the description of the uncertainty. This is part of good practice. All analyses should incorporate a sensitivity analysis, that is, an examination of the way results depend on their assumptions.

From a design perspective, we must be most concerned with the robustness of our design choices. Note that the question is not: Would reasonable changes in our assumption change our valuation? Of course they would. We already know that, in an uncertain world, we do not know what the actual value of our designs will be. The key question is: Would reasonable changes in our assumptions lead us to change our preferred designs? Are we making the right choices? If design A is preferred to design B for a wide range of assumptions, then we can choose it confidently. However, if the choice between these designs depends critically on the assumptions made, then we have to be more careful. In any case, the object of the sensitivity analysis is to check that our preferred design is in fact better across reasonable assumptions.

Difference curves provide a convenient way to present the results of a sensitivity analysis. Because they display the effect of changes, they immediately provide the kind of sensitivity information we can use. For example, the analysis of the parking garage assumed that the uncertainty involved a range of 50 percent variation up or down around the base case demand parameters. We can examine the sensitivity of the results, specifically the conclusion that the flexible four-level design was best overall, by redoing the analysis with different assumptions. Figure 6.16 shows the results for examining the difference in results if we assume that the range of variation is 25 percent or 75 percent instead of 50 percent.

We interpret these results by seeing to what extent changes in assumptions about uncertainty would affect the decision. In figure 6.16, we see that flexible design leads to a net increase in ENPV as we move from the fixed five-level design to the flexible four-level design over the entire range of possible uncertainty examined (i.e., from ±25 percent to ±75 percent). Whereas the evaluation leads to different values for different assumptions, the difference in ENPV is strongly positive in all these cases (from about $1.9 to $2.2 million). We thus see that the selection of the flexible design can be sustained, based on ENPV, even though, given uncertainties about the range of variation in demand, we might not know precisely what the ENPV might be.

Figure 6.16 also illustrates the fundamental fact that the average value of flexibility increases as uncertainty does. As the demand deviation increases from 25 to 50 to 75 percent, the difference in ENPV between the flexible and the rigid design increases. This reflects a basic truth: The greater the uncertainty, the greater the value of flexibility. Put another way, uncertainty is the great driver of the value of flexibility. This is why

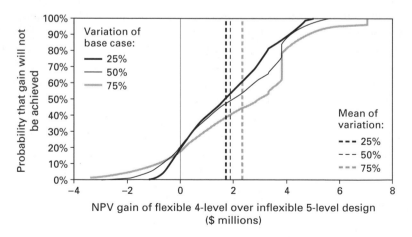

Figure 6.16
Sensitivity of value of flexible design to level of uncertainty (base case compared with higher and lower levels).

flexibility is so important in the design of large-scale, long-term projects: These are the most uncertain projects, and thus the ones where flexibility is most desirable.

In the case of the parking garage, the expected value of flexibility increases from $1.9 million for 25 percent variation to $2.2 million for 75 percent variation. In other words, if demand is more uncertain than anticipated in the original analysis, the flexible design will be worth more on average. The intuitive reason for this is that the flexible four-level structure exploits demand upside but avoids exposure to demand downside. Therefore, when the variation around the base case demand is larger, the upside is more likely, and we are more likely to use the flexibility.

Example Application: Development of a Deep-Water Oil Field

This example concerns a real-world case developed collaboratively by Lin (2009), MIT faculty members, and a major oil company. We developed this material to demonstrate the value of flexible designs for major real-world projects. The application was very successful: It demonstrated a 78 percent increase in expected value for the project while reducing initial capital expenditures (Capex) by about 20 percent.

The Physical Situation

The oil field lies in deep water off the coast of Angola. The field appears to be a series of reservoirs located across a wide area. It is relatively isolated, in that it is not close to other former or currently operating fields, and it does not have access to existing underwater pipelines or storage facilities. The development of the field will thus be self-contained. The company will extract oil and gas using one or more platforms. These will support facilities for primary treatment of the product (such as extracting water and compressing gas) and will store it until it is delivered to ships that will carry it to refineries.

It is impossible to know the effective size of the field until long after the original platform is built and production starts. This is because the effective size depends on factors that cannot be fully determined until production is well under way. Indeed the quantity of oil that it is economical to access depends on the permeability of the deposits, the detailed nature of fractures in the field (which alter the pressure fields that drive the flow of oil and gas), the viscosity of the crude and the amount of gas and water in the deposit. Moreover, what is economical

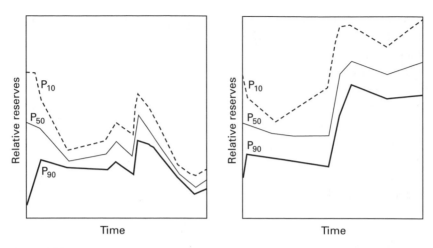

Figure 6.17
Example variations in the best estimates of oil that the operator can extract economically from a field between the time of appraisal and when pumping starts. Notice that in both cases the most likely P_{50} estimate of recoverable reserves at "first oil" differs from the original estimate by a factor of two: half as much for the left-hand field, double the side for the field at the right.
Source: BP sources from Lin, 2009.

to extract obviously depends on product price. If the oil price is high, it will be worthwhile to spend more effort pumping fluid into the field to increase pressure and the amount extracted. As figure 6.17 indicates, estimates of the quantity of recoverable oil in a field can be very uncertain.

Base Case Design Process

A base case design was available at the beginning of the effort to develop flexible design alternatives. The design teams within the oil company had developed it along conventional lines, standard in the major oil companies we have worked with.

The design teams used deterministic forecasts to develop the base case design. Although the geological side of the company estimated the quantity of recoverable product[11] within a broad range (the P_{10} and P_{90} values were about ±50 percent around the P_{50} value, as in figure 6.17), the design team focused on a single value. Similarly, the design team stuck with the instructions developed by upper management to assume that the price of oil would be fixed over the life of the project, at the level management had accepted (at that time) to be the long-term average price of oil.[12]

These deterministic criteria for design naturally burdened the design process with a substantial handicap. Most obviously, they led to erroneous assessments of expected value, as we discuss in chapters 1 and 3 and in appendix A on the flaw of averages. Specifically, it channeled the design toward a fixed capacity, which would cut off benefits from flows that might be higher than anticipated and fail to protect from the losses if the flows were less than anticipated. Evaluations based on most likely estimates will necessarily give incorrect answers because they ignore the reality that constraints shape actual returns. Furthermore, this tunnel vision fixation on average values leads design teams to ignore interesting potential. If we assume that the price of oil will always be $50 a barrel, then of course we ignore opportunities that might be profitable at $60, $70, or $80 a barrel, prices that have occurred and are likely to do so again.

The fact that the base case design used deterministic estimates of recoverable oil and of oil price, which in fact are so uncertain, provided great assurance that flexible designs could provide far greater expected value. Indeed, because the long-term prospects were so uncertain, and because uncertainty drives the value of flexibility, the case study team had every reason to believe that flexible designs would be enormously valuable—as they turned out to be.

Development of Screening Model
The design team for the oil company worked with a highly detailed representation of the system. This "oil-and-gas" model consists of three major parts:

• *A geologic model* describing and optimizing the flow of oil and gas through the oil fields under different designs and operating strategies. This is highly complex, nonlinear, and extensive. It takes significant time to run over the lifetime of the oil field.

• *A model of the man-made developments*, including platforms, wells, sub-sea pipelines, and storage facilities. This is also highly complex, representing hundreds of significant subsystems that designers can combine in many ways. It includes optimization routines to automate the selection of appropriate subsystems into a coherent design. This model is also highly complex, nonlinear, and extensive—and takes significant time to run.

• *An economic model* representing the cash flows of expenses and revenues and ultimately calculating the NPV of a design operating over the lifetime of the project. This is comparatively simple and direct.

To analyze a particular design, analysts had to hook all three models together. A single pass through this apparatus could have taken days. The highly detailed model was inappropriate for a Monte Carlo simulation. In this situation, as is often the case for significant projects, and as we describe in chapter 5, it was necessary to develop a simplified model that the simulation could run quickly, within minutes.

The case study team thus developed a screening model. It was "mid-fidelity," in contrast to the "high-fidelity" detailed "oil-and-gas" model the design team for the oil company normally used. As indicated in chapter 5, the development of the screening model started with a response surface for the high-fidelity model, that is, a representation of its overall behavior. With suitable fine-tuning, we ended up with an acceptable mid-fidelity screening model that enabled us to run Monte Carlo simulations in minutes.

Flexible Designs

The base case design featured a large oil platform optimized for the assumed size of the oil field and product price. It performed well under these conditions. However, if quantities, oil, and prices were low, the design was a poor investment because it was much too big. If quantities of oil and prices were high, the base case design was too small to take full advantage of these opportunities. In short, the optimized base case design performed poorly over a wide range of possible scenarios. This is a prevalent phenomenon. Solutions optimized for one particular future scenario will typically perform poorly when other futures occur. Conversely, solutions that appear suboptimal will often perform much more robustly across a range of futures.

The initial flexible designs targeted the weaknesses of the base case design. They used smaller modules (which limited the downside consequences) that the company could expand (to take advantage of good opportunities, should they occur). Some of the alternative architectures did not increase value—the modules were too small to take advantage of economies of scale in construction. However, experiments with different designs led to a configuration that increased ENPV by almost 50 percent (strategy 5 in figure 6.18).

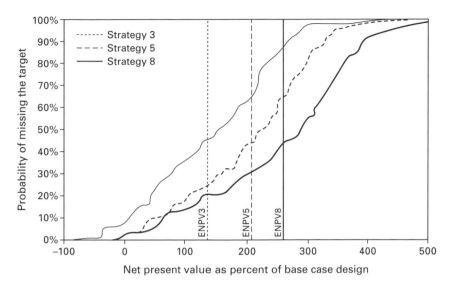

Figure 6.18
Target curves for alternative designs for the development of the deep-water oil field.
Source: adapted from Lin, 2009 (repeat of figure 6.5).

Additional interesting opportunities for flexible design emerged from detailed technical conversations with the oil company. As frequently happens, understanding and working with the engineering details enables designers to uncover particularly clever designs. In this case, a flexible design of the undersea pipelines between several fields significantly improved the overall performance. This is because much of the profitability of an oil field depends on the flows of crude that operators can achieve. If the product is too viscous, it may effectively cap production. As can be imagined, the quality of the oil that operators may recover is uncertain. In this situation, a network of interconnecting undersea pipelines (technically known as tiebacks) provides flexibility to control the quality of oil in the pipelines by mixing flows from several wells. This operational flexibility, combined with flexibility in constructed capacity, led to a 78 percent increase in ENPV over the base case (strategy 8 in figure 6.18).

Multidimensional Evaluation

Although strategy 8 looks like a winner from the perspective of maximizing value, the ENPV does not tell the whole story. The team therefore conducted several multidimensional comparisons, such as the one in

Table 6.3
Multidimensional comparison of selected design strategies for development of an oil field

	Values, as percent of Base Case ENPV or Capex							
	NPV			Capex			Expected	
Strategy	ENPV	Min.	Max.	Expected	Min.	Max.	Reserves	Tiebacks
Base Case	146	–99	400	**100**	**100**	**100**	100	0
5	204	–83	463	131	93	161	146	3.8
8	**257**	**–71**	**578**	167	115	193	**178**	6.3

Best performance in each category highlighted in bold.
Source: adapted from Lin (2009).

table 6.3. This shows that although the expected value of the flexible design is indeed 78 percent better than the base case design (257 compared with 146), it also costs more to initiate (a minimum of 115 compared with 100). The flexible design is also likely to end up costing more over its lifetime (167 to 100), but a large part of these capital investments would come once the upside potential had been validated through experience and so would be much less risky than the initial investment. Considering all these features, the most flexible strategy may indeed be the preferred choice.

Sensitivity Analysis

Analysts should of course ask themselves whether their conclusions are simply a product of some key assumption. In this spirit, the design team examined the sensitivity of the choice of strategy 8 to various assumptions. Table 6.4 shows the results of the sensitivity analysis with respect to the cost of acquiring flexibility, in particular by investing extra capital in the sub-sea tiebacks. The original analysis assumed that the extra cost of the tiebacks would be about 4 percent of the Capex of the base case, as indicated by the bold face in table 6.4. As expected, strategy 8 looks better and better as the cost of flexibility decreases. However, if the cost of flexibility becomes too expensive, strategy 8 would underperform the inflexible base case. If the flexibility cost 20 percent of the base case, strategy 8 delivers lower ENPV (91 compared with 100) at a higher cost.

Such sensitivity analyses are useful to decision makers because they indicate the range over which a choice, such as for strategy 8 over the base case, might be desirable. In this case, for example, one might con-

Table 6.4
Example sensitivity analysis for development of oil field: effect of different costs of flexibility on the value of the strategy 8 compared to base case.

Metric (as percent of base case ENPV or Capex)	Strategy 8							Base case
	Flexibility cost (percent of case Capex)							
	0	2	**4**	8	12	16	20	
ENPV	198	188	**177**	155	134	112	91	100
Min ENPV	43	33	**22**	2	–19	–40	–61	–66
Max ENPV	328	327	**335**	315	295	274	253	251
Expected Capex	169	173	**177**	185	192	200	207	100

The original analysis assumed that flexibility cost was about 4 percent of Base Case Capex. Best performance in each category is highlighted in bold.
Source: adapted from Lin (2009).

clude that strategy 8 would be a preferable choice as long as flexibility cost no more than 12 percent of base case Capex. If we believed that to be true, then we could confidently choose strategy 8 even if we were unsure of the exact cost of flexibility.

Take Away

This chapter presents and demonstrates by example a three-step process for evaluating and choosing designs in a context that recognizes uncertainty and examines flexible design alternatives.

The first step consists of using the Monte Carlo simulation to determine the distribution of possible outcomes associated with any possible design. The application of the simulation proceeds along usual lines. However, analysts have to take special steps to evaluate flexible designs properly. Specifically, they have to insert "rules for exercising flexibility" into their model of the system. These indicate when management should implement flexibility associated with a flexible design. This step develops distributions of performance of the alternatives, conveniently presented as target curves and tables.

The second step involves a multidimensional analysis of the various benefits and costs associated with projects. This process distinguishes the evaluation of designs in an uncertain context from conventional evaluation, which supposes a deterministic context. The process recognizes that project leaders will have to choose between projects using criteria that they cannot usefully collapse into a single metric. At the very least, they will have to balance risks and rewards in several ways. The comparison of alternatives involves comparisons between target curves usefully highlighted by difference curves. Comparative tables are also most helpful, especially for incorporating metrics, such as capital costs (Capex) that are not part of the target curves.

The third step validates the evaluation and choice by detailed analysis of the possible timing of investments in implementing flexibility. It also examines the sensitivity of the choice of a design to the upside and downside of a project, to possible regret, and to assumptions about the nature of the uncertainty.

This chapter illustrates these steps with specific applications. It presents detailed analysis of the simple case of the parking garage. This example has the merit of being easy to understand intuitively. To demonstrate how the process can be successfully applied to major projects, the chapter ends with a thumbnail sketch of the results and benefits of this approach to the design of the development of a major deep-water oil field.

Phase 4: Implementing Flexibility

It's not enough to have an idea; you have to know how to move the furniture around.
—Senator Timothy Wirtz

It is not enough to design flexibility into the system; designers should foster conditions that will enable future managers to use this asset when desirable to do so. If designers do not facilitate the implementation of flexibility as needed, they will reduce its value and waste much of their effort to create the capability. Designers thus have the responsibility to ensure that the flexibility they develop will continue to be available. They need to do their best to keep the flexibility "alive."

Maintaining the capability to implement flexibility is more than a technological issue; it is a social process. The physical capability to make use of the flexibility designed into a system is unlikely to disappear. The extra strength built into a building to allow for the construction of more floors will continue to be available indefinitely. However, a wide range of institutional, regulatory, or political developments may easily prevent the owners of the building from using this strength. For instance, new zoning codes might change the allowable height for a building on its property. New safety regulations might require unanticipated space for emergency exits for all new construction. Bankers might not be willing to finance the current owners. To create useful flexibility, the design process needs to consider and deal with such matters. Designers need to extend their thinking beyond the purely technical aspects of a project.

The process of maintaining flexibility extends the design period well into the future. Just as the logic of dealing with uncertainty impels us to design for physical developments that we can implement as needs or opportunities arise, it also impels us to think through the implementation of these developments over time. In this context, the design process does not stop when it delivers the original plans to the developers; it continues so long as design flexibilities are there for possible implementation.

In this chapter, we provide guidance on how to make sure that managers can implement the flexibility designed into a system as and when desirable. By way of motivation, we first catalog and illustrate common phenomena that prevent future managers from making good use of flexibility in design. We then propose both initial preventive and ongoing operational actions to preserve the capability to implement flexibility. The chapter concludes with an illustration of an organization that has put into practice the principles we recommend here.

Common Obstacles to Implementation

There are many obstacles to making good use of design flexibility. Although we can organize these in several ways, it seems useful to consider five general categories:

• *Ignorance*   Future managers forget about or otherwise ignore that flexibility exists.

• *Inattention*   Nobody monitors or pays attention to the circumstances that would trigger the appropriate use of the flexibility, and managers miss good opportunities to use this asset.

• *Failure to plan*   The design process simply did not think through what needed to be done to implement flexibility, and thus created insurmountable obstacles.

• *Stakeholder block*   Groups using or affected by the system think that the implementation would hurt them and manage to block the use of the flexibility.

• *External developments*   Regulatory, political, or other developments eliminate or otherwise constrain the right to implement the flexibility.

Let's consider each of these.

Ignorance
Future owners or managers of a system may simply not understand the flexibility designed into a system if they were not part of the original

Box 7.1
Ignorance of flexibility: Parking garage in a commercial center

The multilevel garage in a major shopping mall in Britain illustrates how ignorance can prevent the implementation of flexibility designed into the system. When it would have made sense to expand the structure, some 5 years after its opening when the mall was doing well, the owners did nothing because they had no idea of what they could do.

Indeed the developers had sold the garage to a new owner, as is usual. Developers are in the business of taking greater risks with the hope of greater rewards, and they routinely sell off properties once they have become profitable. In this case, the new owner was an investment trust with a large portfolio containing a wide range of properties. They were financial experts, not designers, and they did not think of design capabilities.

The capability to expand the garage was not self-evident. Although the engineers had carefully built the facility with the capacity to add on several levels, this was not visible. The foundations and columns were much stronger than needed for the original structure, but their strength was hidden. The construction process had obviously buried the foundations and carefully encased the steel in the columns in protective concrete. So the new owner could not see the capability for expansion when conducting the due diligence inspection of the property. It was a classic case of "out of sight, out of mind."

design process. This is the professional counterpart to our personal experience: How many of us truly understand the full capabilities of our laptops or cell phones? Obviously, if owners do not understand what they have, they will not think about using it.

In a similar vein, the process for managing the system may easily lose sight of the flexibility. The people responsible for this knowledge may retire or move on to different responsibilities. Most commonly, the system will have become the responsibility of new decision makers through a sale, merger, or change in government. Box 7.1 provides a typical example.

Inattention

To implement flexibility well, systems owners have to do it at a suitable time. This means that they have to pay attention to developments. When the flexibility is associated with routine processes, this may be easy. For example, equipment manufacturers with flexible production lines easily adjust these facilities according to the orders that come in.

Similarly, airport terminals with "swing" gates allocate them to international or domestic passengers over the day as the mix of traffic changes.[1] In such cases, the signal that it is time to implement the flexibility is almost automatic and comes directly to the people in charge of the facility. The airport or plant managers sense the demand and can make changes.

Difficulties arise when the flexibility addresses uncertainties that are not routine. They are compounded when the people or departments that sense the opportunity to exercise flexibility are not those capable of doing so. Consider a fleet of communication satellites designed with the flexibility to reposition themselves to serve different needs or markets. The sales department might know that it would be nice to have some additional capability in an emerging market, but it has neither the ability to implement the flexibility to obtain the new capacity nor even the knowledge that the flexibility exists. Similarly with the development of oil platforms: The production business units may want new capabilities, but they do not understand that the design is ready for it and lack the ability to get project designers to pay attention to their desires.[2] Because of a lack of attention in the right places to the circumstances that call for the implementation of flexibility, managers may not be able to implement it.

Failure to Plan

Implementing any significant form of design flexibility takes effort. It is not just a matter of flipping a switch (or, as in the case of financial options, of notifying a stockbroker). Managers have to obtain the budget for the project, coordinate with existing users and stakeholders of the system, obtain local and possible international permissions, secure contractors or others to execute the operation, and so on. Implementing flexibility requires careful planning. Managers need to know how "to get from *here* to *there*."

The problem is that the designers of the technical system often fail to think through what it takes to implement their flexibility. They need to develop an implementation plan and take the corresponding steps to put the appropriate capabilities in place. Sometimes the failure to plan is glaringly obvious, as with building complexes in which the heating and cooling plants are placed as close as possible—and right in the way of any possible expansion, making it impossible.[3] More generally, the difficulties can be subtle, as box 7.2 illustrates.

Box 7.2
Failure to plan: Newcastle hospital

The structural design for the Royal Victoria Infirmary in Newcastle-upon-Tyne provided the strength to add extra floors for additional future wards. This gave the owners the physical ability to adapt the hospital to future needs as required.[9]

However, the owners were not able to implement this flexibility because of a seemingly trivial administrative detail. The hospital authorities developed the hospital under the UK PFI. This procurement mechanism uses a private-sector company to own, finance, build, and maintain infrastructure—in this case, the hospital. The company is a "special purpose vehicle" created to develop the facility and rent it to the public client under a long-term agreement. The special purpose vehicle for the Newcastle hospital financed the development with bonds. These were risk-rated based on the original plans, which included the flexibility to expand the hospital vertically to provide extra wards. However, it turned out that the hospital's desire for extra space did not come from a shortage of wards but from the need to relocate offices, which would ideally fit into an additional floor. The difficulty was that the original plans did not include this possibility. Although this seemed like a trivial difference, and the expansion for offices would have been easier and cheaper than for wards, this small detail would have triggered the expensive need to re-rate the bonds and refinance the project. The owner company deemed this unprofitable. The hospital could not use the technically available flexibility because of these small details omitted from the original plans.

Stakeholder Block

Many stakeholders participate in the execution of any important project. Their acceptance and cooperation may be essential to the implementation of flexibility. Because they often block projects, managers must deal with these interest groups.

Stakeholders internal to the organization may be significant. Groups within an organization may resist some action that, although good for the organization, adversely affects their standing or benefits. Consider the situation of the organization running the fleet of communication satellites: The group in charge of the operation of the satellites may resist requests for repositioning the satellites to expand markets. Why? Perhaps because the company measures the performance of the operating department on their costs, so this group does not want to spend whatever is necessary to implement the flexibility, which almost certainly is not in their normal budget. Likewise, operators may be concerned about

Box 7.3
Stakeholder block: Houston metro

The designers of the Houston metro carefully arranged for the possibility of constructing an office building and parking garage near one of the metro stations close to the Texas Medical Center. They estimated correctly that the location, next to the hospital complex, was ideally convenient and that the development could be highly profitable.

The hospitals also recognized the value and built medical offices and parking garages in and around the Texas Medical Center. These facilities were profitable and became a major source of funding for the hospitals. The hospitals thus became important stakeholders in any future development around the metro station, which would become competition and undercut their finances.

When developers tried to implement the flexibility designed around the metro station, they found it impossible to do so. The hospitals appear to have pleaded with the political authorities, asking them not to create competition that would threaten their finances and their ability to serve the community. In the end, the hospitals corralled enough political support to prevent the metro system, itself governed by a political process, from going ahead with the project.[10]

This case illustrates the fact that stakeholders affected by the implementation of some flexibility can often organize to prevent such a development.

possible technical difficulties and would prefer not to mess with a system that is operating well. In any case, their cooperation will be essential to the implementation of the flexibility of repositioning the satellites.

External stakeholders are also often significant. They may also be more difficult to work with than internal stakeholders. Within a company or organization, there may be a decision maker who can resolve internal disputes authoritatively, such as might exist between the sales and operating departments of the organization operating communication satellites. External stakeholders by definition are not subject to such power. System managers will have to negotiate with them somehow or use the political process to implement desired flexibility. Box 7.3 illustrates this issue.

External Developments
Finally, all kinds of external developments can make it impossible or impractical to implement flexibility. For example, new building codes may prohibit what was previously acceptable, as has happened when concern for disabled persons led to major changes in design practice.

System managers may be powerless to prevent changes that can block their use of flexibility. They can, however, monitor developments and plan to implement flexibility (if desirable) before the changes take place. Thus, the staff of the Health Care Service Corporation (HCSC) carefully monitored the evolution of the zoning codes in Chicago to ensure that the HCSC would be able to exploit its flexibility to expand its building vertically, as chapter 1 describes. For them, as for other developers, good working relations with the authorities were essential to their ability to implement planned flexibility.[4]

Initial Preventive Actions

To maximize the likelihood of being able to implement design flexibility, the design process can take both initial preventive and ongoing operational actions. The former leads to the latter.

The initial preventive actions cluster into three major types:

• *Integrated project delivery*   Creating the design with the participation of major stakeholders in the process, thereby uncovering and understanding many of the issues that might eventually be barriers to implementation.

• *Development of game plan*   Carefully thinking through what would be required for future implementation, avoiding the creation of obstacles while laying the groundwork for easy implementation.

• *Preparatory action*   In accord with the game plan, taking actions that increase the potential for implementing the flexibility, should it ever be desirable.

Integrated Project Delivery

Integrated project delivery is a nontraditional process for designing and implementing projects. Project developers are increasingly trying this innovative process, but it is still a novelty. The basic concept is that all the stakeholders involved with the delivery of a project should work together collaboratively and simultaneously. Designers work with manufacturers and prospective operators to create facilities they can build and maintain easily, and these technical professionals also work with forecasters and financial institutions to plan coherent phasing and flexibility of the project.[5]

Integrated project delivery contrasts with the traditional process, in which distinct professional teams work independently. The stereotypical version of normal project delivery is that forecasters or others develop requirements for a system; the process passes these on to designers who translate these specifications into drawings; these plans are then turned over to manufacturers who try to build the system as best as they can; and so on. As the description indicates, this process limits the scope for flexibility. For example, the manufacturer has little ability to influence the concept of the design, which engineers have fixed by the time it arrives for implementation. Moreover, the traditional process does not give participants much opportunity to understand each other's perspective and concerns, so they have little chance of anticipating and dealing with issues that may hinder eventual implementation of flexibility.

An integrated process of project delivery is useful because it enables participants to anticipate and avoid issues. For example, if the designers of the Royal Victoria Infirmary in Newcastle (see box 7.2) had been in touch with the people organizing the financial arrangements, they could have made sure that the loan documents included a more flexible description of possible evolution of the project and thus facilitated onward financing. Vice versa, if the financial process had been aware of the clever way the designers were planning the hospital, they could have taken the initiative to make sure this was clear to prospective lending agencies.

Development of Game Plan

The development of a game plan for eventual implementation is basic to successful implementation. The idea is conceptually simple: Designers should lay out the steps that managers would need to take to implement each particular form of flexibility and to anticipate how best to carry it out. The difficulty lies in effective implementation.

A good game plan will correctly anticipate what tasks are necessary for implementation. Unless the design process consults closely with the stakeholders in the project, they are likely to miss important details. Thus, with the Royal Victoria Infirmary: The architects and engineers had doubtless thought through the purely technical construction process for implementing the flexibility to expand the hospital. However, whatever they planned seems to have been wasted because appropriate financing arrangements had not been included in the plan. Thus, some form of integrated project process is essential for the development of a good game plan for future implementation.

Preparatory Action

Following on from the game plan, the design process may anticipate that managers must take immediate actions to facilitate future implementation of flexibility. These actions may be as necessary to success as the technical actions that the design has inserted in the system.

For example, U.S. national aviation authorities (the Federal Aviation Administration [FAA]) have been anticipating the possibility that Chicago will need a major new airport, and they have been working with regional authorities to develop a plan. In this case, they have generally worked out the design and location. However, this is not enough to achieve implementation should that ever be desirable. Therefore, the FAA has also been supporting the purchase of the land that would be needed. Their reasoning is that if they waited to buy the land once they actually decided to build the airport some years in the future, it might have been transformed from farms into suburban developments and be impossible—politically or financially—to acquire.[6] Similarly, Boston and Los Angeles have kept the lightly used airports of Worcester and Palmdale operational to ensure the flexibility to use their capacity if that ever made sense.

Ongoing Operational Actions

The owners and operators of a system need to help sustain the ability to implement flexibility for as long as this may be useful. In so doing, they act as extensions of the initial design process, which we cannot consider to be complete until all possible flexibilities have been used or discarded. Three kinds of actions seem most useful in this regard:

• *Maintaining the right to implement*   Because the ability to implement is often contingent on various legal permissions, it is important to keep these up-to-date and in effect.

• *Maintaining the knowledge to implement*   Effective implementation requires people, and more generally institutions, that understand the nature of the flexibility and know how to proceed when the opportunity is attractive.

• *Monitoring the environment*   This is crucial to knowing when it would be desirable to implement flexibility and to obtain its highest value.

Maintaining Rights

The ability to implement flexibility may require legal or other permissions. These can easily evaporate if not maintained. For example, many European countries now require owners to renew patent rights every year, so companies wanting to maintain the right to produce a product with patent protection must be careful to keep their patents in force.

Likewise, it is frequently the case that planning permission to build a facility will expire if the permit holder does not use it within a specified limit. The watchword in many cases is "use it or lose it."

One of the more effective ways to maintain the right to do something is to keep on doing it from the start. In most regions, it is normal to accept that owners can maintain existing rights to operate almost indefinitely. Someone running a pig farm in an agricultural area that is undergoing suburbanization can expect to carry on the business—those moving in have had to consider this activity. However, a newcomer is unlikely to get permission to start a new pig farm. Similarly, an industrial plant can expect to continue its operations as the city grows around it, even as the political process might stop a new factory. In this spirit, it is essential for Boston and Los Angeles to keep airport operations active at Worcester and Palmdale if they want to maintain the ability to implement the flexibility to expand at those locations. If they ever allow these airports to cease operations, the neighbors will grow accustomed to the closure, and reopening the airports would be highly problematic.

Flexibility can be excessively expensive, and therefore ineffective, through locked-in contracts. Everyone who has built or remodeled a house knows how expensive it is to alter the plans even slightly, for example, by adding an extra window during the construction; there will be no competitive bidding for the change. The UK private finance initiative (PFI) is an interesting example of how such arrangements can hamper the effective use of flexibility. The basic idea of the PFI is that a private-sector company finances, constructs, maintains, and owns critical infrastructure, such as hospitals, schools, or bridges, as part of a long-term rental agreement with the public sector. When unforeseen circumstances arise and changes to this infrastructure would seem appropriate, it is the owner—the private-sector company—who has the right to exercise this flexibility (unless the rental agreement clearly specifies the changes in advance). The private owner can block any changes and will do so unless the public-sector client agrees to favorable terms in a renegotiated rental contract. The public-sector client can thus be locked in, and the flexibility will either not be exercised or be very expensive (see box 7.2).

Maintaining the Knowledge

In parallel, it is important to maintain institutional knowledge about the nature of a flexibility to ensure the ability to implement it in the future. Most fundamentally, it is necessary to know that the flexibility exists (see box 7.1). However, this by itself is generally not enough. It is important to know the crucial details. For example, the Tufts Dental School in Boston was able to exploit the flexibility to expand its building vertically because it could access the plans and knowledge of the local engineers who had originally designed the building a generation earlier. This was a piece of good luck for them, as professional firms merge, principal designers move and retire, and requisite knowledge is generally no longer available.

Organizations planning for the possibility of implementing some flexibility should pay attention to the process of maintaining access to the design knowledge that might eventually be essential. Good documentation is the elementary necessary requirement, but it is not sufficient. People are needed to maintain the institutional memory and the capability to access relevant information. Thus it was with HCSC, the company that doubled the height of its headquarters in Chicago (as described in chapter 1). Its staff kept up close relations with its original architects and suppliers so that when the time came for exercising their vertical flexibility, they had their team in place. If they had not proactively maintained this knowledge base, it is doubtful that they could have implemented the flexibility to expand their building as originally intended.[7]

Maintaining staff continuity can be the essential ingredient for ensuring effective implementation of flexibility. In the case of the HCSC, for example, the process was effective because the same person headed their facilities department from the planning of the original building in Chicago, built with the flexibility to expand vertically, to the completion of the eventual expansion almost 20 years later. Because that person understood the flexibility and maintained the institutional knowledge through his relations with the project architects, the project proceeded smoothly in a way that would have been almost unthinkable without his presence. In this case, as can easily occur elsewhere, the feasibility of a major project depends on just a single person or a few individuals. An important key to implementing flexibility successfully may simply be to maintain professional staff in place—the cost of this insurance can be insignificant compared with the size and value of the project.

Monitoring the Environment

Flexibility has the most value when implemented at the right time. It is therefore important to know when this might be. Organizations considering the possibility of implementing some design flexibility thus need to set up and maintain a process to track the factors that might trigger the effective use of this flexibility. This process needs to have at least two features. It should:

• *Highlight triggers* Conditions that indicate when it is desirable to implement a design flexibility; and

• *Establish useful information flows* To ensure that the information needed flows to the people who can use it to decide whether and when to exercise the flexibility.

The design process will normally have identified the conditions that call for the use of design flexibility. This step is an essential part of the process. To be able to value and thus justify any possible flexibility we might build into our system, we should know when and under what conditions we might use it. In fact, these conditions define the "rules for exercising flexibility" that we need to embed in the Monte Carlo simulations, as indicated in chapter 6. These are the triggers that the organization and its managers need to keep in mind over the life of the system.

The issue is that the design team needs to highlight the conditions for implementing flexibility. It needs to pass the knowledge about these on to those who will be responsible for the design in the future. We need to bring forward the understanding of the triggers that justify the implementation of design flexibility, from the depths of the analysis to the top of the organization that will operate the system in the future. As part of its planning for implementation, the design team needs to create signposts to guide future managers.[8]

Highlighting triggers for the implementation of flexibility may not be easy. Sometimes the conditions justifying the use of flexibility are obvious, and great efforts are not required. In the case of HCSC, which had a need to keep its staff on one campus, the trigger was closely associated with the number of staff and the difficulty of crowding more employees into the existing facilities. As space became tight, the desirability of expansion became clear. However, the conditions favorable for the use of flexibility are generally not obvious. They often combine several factors. The desirability of expanding the capacity of an oil platform, for example, depends on the price of oil, the estimates of the amount recoverable, and the speed at which operators can extract the oil. Moreover, the conditions that would reasonably trigger the implementation of the flexibility are likely to combine in complex ways—very high oil prices might compensate for lower estimates of recoverable oil, for example. Creating and maintaining suitable signposts for future managers can be difficult.

Having good signposts is not sufficient in any case. It is also necessary to make sure that relevant information will flow to the future leaders of a system so they can compare signposts and current information. The essential part consists of confronting the available data and the trigger conditions. This is what is required to create the conditions for thinking about and eventually choosing to take advantage of flexibility in design.

It is necessary to create a process whereby appropriate information is available at the right place. Normally, an organization will have the relevant information available somewhere. The question is whether it is in the right form and the right place. Consider the example of expanding the capacity of an oil platform. We may safely assume that different departments of the oil company will be fully aware of the price of oil, the estimates of recoverable oil in place, and the pressure fields that control the pace of extraction. The difficult part is making sure that such information comes together usefully and becomes available to those in a position to decide whether to implement the design flexibility.

The process of monitoring the environment to know when it might be desirable to implement flexibility is obviously crucial. If we do not have a means of knowing when to act, clearly we are unlikely to do so. Unfortunately, there is no simple way to ensure that the information will be available to the right people in good time. Different approaches will be better in different organizations.

Example Application: Dartmouth-Hitchcock Medical Center

The Dartmouth-Hitchcock Medical Center (DHMC) provides a good example of how to design and implement flexibility. The DHMC built facilities that it can expand and otherwise alter flexibly and set up an organization that ensures timely and effective use of the design flexibility.

The DHMC is located in the State of New Hampshire in the United States. It is a major regional hospital closely associated with Dartmouth, a university with a leading medical school. It has to serve four important stakeholders. In addition to providing surgical facilities and hospital beds for major interventions, it must also cater to medical practices that serve walk-in patients from the region; respond to the clinical needs of around 10,000 students, staff, and dependents of Dartmouth College; and provide special facilities for research carried out under the sponsorship of the U.S. National Institutes of Health, which have their own requirements.

In the 1980s, the DHMC had the opportunity to move from a con-
glomeration of facilities nestled on the university campus to a large new
site. The DHMC took advantage of this opportunity to create a flexible
design that would allow each of their major stakeholders to expand their
facilities as and when needed. Also—and this is the point of this example—the DHMC set up an organization and processes to facilitate effective implementation of the design flexibilities.

Flexible Physical Design

The essence of the design for the DHMC is that facilities for each of the
major stakeholders are spaced along a central spine, which provides for
circulation and a variety of common services throughout the DHMC (see
figure 7.1). The buildings serving each of the major clients extend out at
right angles from the spine. These facilities differ in size and content
according to the different needs of their users. The design kept certain
elements uniform, such as floor heights, to facilitate communication,
utilities, and the like. Because these facilities are essentially independent,
they can expand or change function at their own speed according to their
needs and financial capabilities. For example, at one point, the DHMC
added extra in-patient beds (at the far left of figure 7.1). Most recently,
toward the right side of the plan, the research group added floors to its
building.

Notice also that the design locates the utility plant (providing heating
and air conditioning for the DHMC) far away from the medical buildings. This is unusual. Common practice places the utility plant close
to the buildings it serves to minimize the cost of ductwork and to reduce
the energy losses along the length of these pipes. In this case, however,
the design recognized that a remote location was an essential part of the
flexible design. If the utility plant were next to the medical buildings, it
would limit their possibilities for future expansion. Moreover, the original design made the utility plant much larger than it had to be to serve
the initial facilities. The idea was to facilitate eventual expansion as and
when needed.

Design for Implementation

The DHMC provides an excellent example of how it is possible to design
for implementation of flexibility. In their case, they created a separate
organization to manage and develop the facilities. This group is a special-
purpose consortium of the major stakeholders. Its day-to-day role is to
manage common utilities and services, such as parking. The consortium

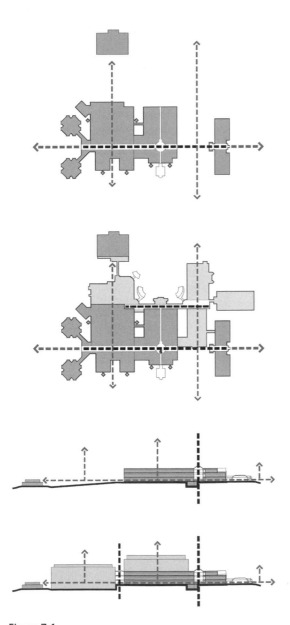

Figure 7.1
Plan and elevation views of Dartmouth-Hitchcock Medical Center. The plans show how the original design enabled independent expansion, both vertically and horizontally at right angles to the spine, of the facilities serving the several major stakeholders.
Source: © 2010 Shepley Bulfinch architects, Boston.

has a long-term role to maintain relationships with architects and other designers, to oversee all construction, and, in general, to be the common agency for construction management for each of the four major stakeholders. In turn, representatives of the stakeholders meet regularly with the facilities group to keep abreast of ongoing issues and the various aspirations for further developments.

The DHMC facilities group provides a role model for how to achieve each of this chapter's recommendations for implementation:

• *Integrated project delivery*   The DHMC group provided the central node to coordinate and mediate the various desires and issues of stakeholders, designers, construction crews, and local and regional planning authorities.

• *Development of game plan*   The DHMC had clearly thought through how it would proceed to implement the development of the hospital facilities in its ever-changing technological and regulatory environment. They made sure the physical plan enabled separate expansions according to the diverse needs and opportunities of the stakeholders and—by establishing the separate facilities group—gave themselves the ability to monitor and respond to events coherently and consistently.

• *Preparatory action*   The DHMC not only ensured that the original utility plant was located remotely, but also carefully sited subsequent supporting facilities. For example, it built structured parking with restraint, giving great emphasis to preserving ground-level parking near the buildings to maintain the ability to develop the site easily.

• *Managing rights*   The original design could be described as a small set of buildings surrounded by a large area of ground-level parking. Creating this large surrounding space required DHMC to clear and grade much more space than they might otherwise have done. However, the benefit of clearing this space and using it from the start is that they have effectively established the right to use the cleared area. They will not have to face public pressure to maintain the wooded areas, as they might have had to if they had only cleared a smaller area.

• *Maintaining the knowledge*   The DHMC facilities group has managed to maintain knowledge of flexibility by holding together a consistent team over more than two decades. Their facilities manager and her team have been in place a long time and have been working with the same architects over that period.

• *Monitoring the environment*   Through their regular meetings with the stakeholders, the DHMC facilities team manages to stay up-to-date with stakeholders' ambitions for expansion, financing possibilities, and the timing of possible implementations of the flexibility designed into the original system.

Take Away

Flexibility that we cannot implement when needed has little value, so it is essential for designers to pay close attention to how future system managers could implement any flexibility designed into a project.

Thinking about implementation stretches conventional concepts of design. This is because many of the obstacles that may have to be overcome are social, involve legal requirements, economics and financing, institutional rivalries, and, of course, politics. Moreover, the implementation phase for any design flexibility may last years, often a decade or more. This is not the common view of design practice. Yet if effective design involves getting things done, as we believe it should, then design for flexibility extends practice both over time and into social issues.

To ensure the possibility of effective implementation, designers should pay attention to both the immediate design process and its extension over time. In the immediate, they should encourage collaborative consultation between stakeholders, in some form of integrated design process, and develop a game plan for implementation. Over the life of the project, or as long as design flexibilities may be available, it is important to maintain the right to exercise flexibility, establish signposts that indicate when flexibility should be implemented, and establish processes that will ensure that timely, relevant information flows to the system leaders who have the authority to implement the flexibility. Good designers can do this, although it sounds difficult. Good designs teams are successful at implementation, as the example shows.

8 Epilog: The Way Ahead

The Context

Uncertainty about the future has always been a concern in long-term projects, but the type of uncertainty these systems face is undergoing a fundamental change. Traditionally, the main risk of long-term technology and infrastructure projects came from shocks, such as wars or natural catastrophes. These events are unpredictable. The design response to this type of uncertainty has been to build in safety margins to ensure safe operation, as far as possible, when such events occur. The design process generally neglects the economic effect of the uncertainty. This is understandable, because wars or natural catastrophes will normally have such a drastic effect on the social world within which the system operates that the economics of the system is a minor concern. However, the world has changed.

Economic uncertainty now complements and arguably surpasses the traditional political and natural uncertainty. The pace at which commercial realities change today, driven by the information revolution and globalization, is unprecedented. The tremendous rate of technological advance further exacerbates economic uncertainty. The effect of technology, however, is unpredictable. Some technologies, like the Internet and mobile telephony, take the world by storm. Others, like biotechnology, fail to live up to their commercial promise, at least in the short term. This is a new world for the leaders and designers of large-scale engineering systems.

The Promise

Flexibility in design opens the door to significant improvements in the value of projects. In the highly uncertain world in which we live, where surprises almost invariably invalidate the forecasts of future developments, flexibility in design enables system managers and owners to adapt their products and projects to situations as they actually are. They can exploit their flexibility to avoid downside losses and take advantage of upside opportunities. Case after case indicates that the advantages of flexibility in design are highly significant.

Both theoretical and practical grounds lead us to recognize that we need to look for and implement suitable flexibility in design. The best flexible solutions may be difficult to find. The calculations involving possible distributions of outcomes are both more extensive than usual and outside of usual practice. Implementing design flexibility at the right time and place also requires special effort. Yet the overall rewards appear great. First adopters of flexibility in design will thus gain competitive advantage. That is the promise of flexibility in design.

The Current Situation

Unfortunately for engineering practice, few companies or organizations routinely create flexible designs. The standard approach, embedded in established practice and codified in systems engineering manuals, is to design to fixed specifications or fixed best estimates of future conditions. This conventional practice is doubly unfortunate: These procedures not only fail to deliver the higher value that can be achieved with flexible designs, but also—due to the flaw of averages—tend to give misleading and erroneous appraisals of value.

Changing established practice is always difficult. Most people stick with what they know, with what education and professional and personal socialization has imprinted on them. Organizations tend to be particularly resistant to change. Their leaders typically rise to the top because they are particularly successful in the established ways of doing things, are naturally proud of what they have achieved, and are reluctant to leave what has worked so well for them. Organizations themselves encourage conformity and resist change. This is particularly so in large bureaucratic organizations, such as those that lead the development of large-scale technology systems.

There is a kernel of truth in the saying that "science proceeds funeral by funeral." Major change in established paradigms often requires the passing of the old guard and its replacement by the next generation. With this prospect, we can imagine that we are still a generation away from the widespread adoption of flexibility in design.

The Opportunity

The prospect that the concept of flexibility in design will take a long time to be widely accepted by the design professionals gives us a major opportunity. Innovators—individuals or organizations—who adopt this approach will be able to develop and offer superior designs that offer significant improvements in value without widespread competition. Early adopters of this approach will have a competitive advantage over competitors who find it difficult to change their established ways.

The Work Ahead

The procedures for creating effective flexibility in design are not well established. Many of those we describe in this book derive from recent doctoral work at advanced research universities. They are far from definitive. Although they demonstrate promise, they have not been extensively tested in real applications. We have much work to do before we have reasonably definitive agreement on the details of how best to proceed.

We need to continue research to improve our analytic procedures. We need to look for better, more efficient ways to identify good opportunities for flexibility, to improve on our procedures for analyzing potential flexible designs in different situations, and to develop a better understanding of which procedures work best in which situations.

We also need to carry out extensive tests of the approach in realistic situations. We need to proof-test specific procedures in practice, identify and correct their flaws, and otherwise improve them. We need well-documented cases of both what works and what doesn't. We have already assembled a large number of applications in many fields, and we have posted them on the Web site that supports this book (http://mitpress.mit.edu/flexibility). We welcome thoughtful contributions to this inventory from industry and academia. We hope these examples will help us all develop a better understanding of how we can best develop and implement flexibility in design.

Appendix A: Flaw of Averages

The term "Flaw of Averages" refers to the widespread—but mistaken—assumption that evaluating a project around average conditions gives a correct result. This way of thinking is wrong except in the few, exceptional cases when all the relevant relationships are linear. Following Sam Savage's suggestion,[1] this error is the "Flaw" of Averages to contrast with the phrase referring to a "law" of averages.

The Flaw of Averages can be the source of significant loss of potential value for the design for any project. The rationale for this fact is straightforward:

• A focus on an "average" or most probable situation inevitably implies the neglect of the extreme conditions, the real risks and opportunities associated with a project.

• Therefore, a design based on average possibilities inherently neglects to build in any insurance against the possible losses in value and fails to enable the possibility of taking advantage of good situations.

Designs based on the Flaw of Averages are systematically vulnerable to losses that designers could have avoided and miss out on gains they could have achieved. The Flaw of Averages is an obstacle to maximizing project value.

The Flaw of Averages is a significant source of loss of potential value in the development of engineering systems in general. This is because the standard processes base the design of major engineering projects and infrastructure on some form of base-case assumption. For example, top management in the mining and petroleum industries have routinely instructed design teams to base their projects on some fixed estimate of future prices for the product. Likewise, the designers of new military systems normally must follow "requirements" that committees of generals, admirals, and their staff have specified. The usual process of conceiving, planning, and designing for technological systems fixes on some specific design parameters—in short is based on the Flaw of Averages—and thus is unable to maximize the potential value of the projects.

We definitely need to avoid the Flaw of Averages. Because it is deeply ingrained in the standard process for the planning, design, and choice of major projects, this task requires special efforts. The rewards are great, however. The organizations that manage to think and act outside of the box of standard practice will have great competitive advantages over those that do not recognize and avoid the Flaw of Averages.

To appreciate the difficulties of getting rid of the Flaw of Averages, it is useful to understand both why this problem has been so ingrained in the design of technological projects and how it arises. The rest of this appendix deals with these issues.

Current Design Process for Major Projects Focuses on Fixed Parameters

The standard practice for the planning and delivery of major projects focuses on designing around average estimates of major parameters. For example, although oil companies know that the price of a barrel of oil fluctuates enormously (between 1990 and 2010, it ranged from about $15 to $150 per barrel), they regularly design and evaluate their projects worldwide based on a steady, long-term price (in the early 2000s, this was about $50 per barrel of oil). Similarly, designers of automobile plants, highways and airports, hospitals and schools, space missions and other systems routinely design around single forecasts for planned developments.

Complementarily, the standard guidelines for system design instruct practitioners to identify future "requirements."[2] Superficially, this makes sense: It is clearly important to know what one is trying to design. However, this directive is deeply flawed: It presumes that future requirements will be the same as those originally assumed. As the experience with GPS demonstrates, requirements can change dramatically (see box A.1). Designers need to get away from fixed requirements. They need to identify possible future scenarios, the ranges of possible demands on and for their systems.

The practice of designing to fixed parameters has a historical rationale. Creating a complex design for any single set of parameters is a most demanding, time-consuming activity. Before cheap high-speed computers were readily available, it was not realistic to think of repeating this

Box A.1
Changing requirements: Global Positioning System

> The designers of the original satellite-based Global Positioning System (GPS) worked with purely military requirements. They focused on military objectives, such as guiding intercontinental missiles. They were enormously successful in meeting these specifications.
>
> GPS has also become a major public success. Satellite navigation now replaces radar in many forms of air traffic control. It is also a common consumer convenience embedded in cell phones, car navigation systems, and many other popular devices. It provides enormous consumer value and could be very profitable.
>
> However, the original specifications failed to recognize possible commercial requirements. The design thus did not enable a way to charge for services. Thus, the system could not take advantage of the worldwide commercial market and could not benefit from these opportunities. The developers of the system thus lost out on major value that would have been available if they had recognized possible changes in requirements.

task for hundreds if not thousands of possible combinations of variations of design parameters. Originally necessary, the practice of designing to fixed parameters is now ingrained in practice.

Management pressures often reinforce the pattern. In large organizations, top management and financial overseers regularly instruct all the business units to use identical parameters—for example, regarding the price of oil. They do this to establish a common basis for comparing the many projects that will be proposed for corporate approval and funding. A fixed baseline of requirements makes it easier for top management to select projects. However, when the conditions imposed fixed are unrealistic—as so often is the case—these prevent designers from developing systems that could maximize value.[3]

The conventional paradigm of engineering further reinforces the tendency to accept fixed parameters for the design. Engineering schools commonly train engineers to focus on the purely technical aspects of design.[4] A widespread underlying professional view is that economic and social factors are not part of engineering; that although these may have a major effect on the value of a project, they are not suitable topics for engineering curricula. The consequence in practice is the tendency for designers to accept uncritically the economic and social parameters given to them, for example, the forecasts of demands for services or anticipations of the legal or regulatory rules.

In short, the practice of designing to fixed parameters is deeply entrenched in the overall process for developing technological projects. Even though the focus on requirements, most likely futures, or fixed estimates leads to demonstrably incorrect estimates of value, entrenched habits are not likely to change easily. Current and future leaders of the development of technological systems need to make determined efforts to make sure that the Flaw of Averages does not stop them from extracting the best value from their systems.

The Significant Errors

The Flaw of Averages is associated with a simple mathematical proposition. This is that:

The Average of all the possible outcomes associated with uncertain parameters is generally not equal to the Value obtained from using the average value of the parameters.

Formally, this can be expressed as:

$E[f(x)] \neq f[E(x)]$, except when f(x) is linear

This expression is sometimes called Jensen's law.[5] In this formula, f(x) is the function that defines the value of a system for any set of circumstances, x. It links the uncertain input parameters x with the value of the system. In practice, f(x) for a system is not a simple algebraic expression. It is typically some kind of computer model. It may be a business spreadsheet, a set of engineering relationships, or a set of complex, interlinked computer models. It may give results in money or in any other measure of value, for example, the lives saved by new medical technology. $E[f(x)]$ indicates the Expected Value of the system, and $E(x)$ indicates the Expected Value of the parameters x, the average condition under which the system operates.

In simple language, the mathematical proposition means that the answer you get from a realistic description of the effect of uncertain parameters generally differs—often greatly—from the answer you get from using estimates of the average of uncertain parameters.

This proposition may seem counterintuitive. A natural reasoning might be that:

• if one uses an average value of an uncertain parameter,

• then the effects of its upside value will counterbalance the effects of the downside value.

False!

The difficulty is that the effects of the upside and downside values of the parameter generally do not cancel each other out. Mathematically speaking, this is because our models of the system, f(x), are nonlinear. In plain English, the upside and downside effects do not cancel out because actual systems are complex and distort inputs asymmetrically.

The systems behave asymmetrically when their upside and downside effects are not equal. This occurs in three different ways:

- the system response to changes is nonlinear,
- the system response involves some discontinuity, or
- management rationally imposes a discontinuity.

The following examples illustrate these conditions.

The System Response Is Nonlinear

The cost of achieving any outcome for a system (the cars produced, the number of messages carried, etc.) generally varies with its level or quantity. Typically, systems have both initial costs and then production costs. The cost of producing a unit of service therefore is the sum of its production cost and its share of the fixed costs. This means that the costs/unit are typically high for small levels of output and lower for higher levels—at least at the beginning. At some point, a further increase of production volume may require special efforts—such as higher wages for overtime or the use of less productive materials. Put another way, the system may show economies of scale over some range and increased marginal costs and diseconomies of scale elsewhere. The overall picture is thus a unit cost curve similar to figure A.1.

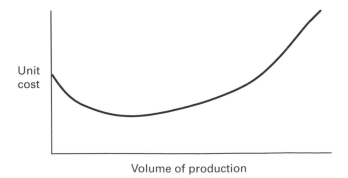

Volume of production

Figure A.1
Typical cost curve for the output of a system.

When the costs—or indeed any other major aspect of the project—do not vary linearly, the Flaw of Averages applies. Consider the case when we have designed a system to operate at the lowest unit cost for an expected scenario, at the bottom of the curve in figure A.1. Then the effects of production levels higher or lower than expected do not average out. It is not the case that the higher cost in one unexpected scenario is offset by lower costs in another unexpected scenario. Whatever unexpected scenario occurs, the unit costs will be higher than anticipated. The average unit cost will then be higher than the unit cost based on expected production volume. Box A.2 illustrates the effect.

The System Response Involves Some Discontinuity

A discontinuity is a special form of nonlinearity. It represents a sharp change in the response of a system. Discontinuities arise for many reasons, for example:

• *The expansion of a project might only occur in large increments.* Airports, for example, can increase capacity by adding runways. When they do so, the quality of service—in terms of expected congestion delays—should jump dramatically.

• *A system may be capacity constrained* and may impose a limit on performance. For instance, revenues from a parking garage with a limited number of spaces will increase as the demand for spaces increases but will then stop increasing once all the spaces are sold. Box A.3 illustrates this case.

Management Rationally Imposes a Discontinuity

Discontinuities often arise from management actions, from outside the properties of the physical project. This happens whenever the system operators decide to make some major decision about the project—to enlarge it or change its function, to close it or otherwise realign its activities. Box A.3 gives an example of how this can happen.

Take Away

Do not be a victim of the Flaw of Averages. Do not value projects or make decisions based on average forecasts. Consider, as best as you practically can, the entire range of possible events and examine the entire distribution of consequences.

Box A.2
Nonlinear system response: Thermal power plant

Consider a regional utility whose production comes from a mix of low-cost hydropower and expensive oil-fired thermal plants. Its average cost per kilowatt hour (kWh) will be lower if Green policies reduce demand and higher if economic growth drives up consumption, as in table A.1. What is the margin of profitability for the utility when it is obliged to sell power to consumers at a fixed price of $0.06/kWh?

If we focus on the average forecast, that is a consumption of 1,000 megawatt hours, then the utility has a margin of $0.01/kwh (= 0.06–0.05) for an overall profit of $10,000. However, if we consider the actual range of possibilities, then we can see that the high costs incurred when the demand is high, boosted by the high volumes of demand, lead to losses that are not compensated on average by the possible higher profitability if demand and prices are low. In this example, the average value is thus actually $1,600 compared with the $10,000 estimate as shown in table A.2. A focus on average conditions thus leads to an entirely misleading assessment of performance. This is an example of the Flaw of Averages.

Table A.1
Cost of supplying levels of electricity

Level of use (megawatt hours)	Probability	Average cost ($/kilowatt hours)
1,200	0.3	0.08
1,000	0.4	0.05
800	0.3	0.04

Table A.2
Actual profitability under uncertainty

Level of use (megawatt hours)	Probability	Cost ($/kwh)	Margin ($/kwh)	Overall profit ($)	Expected profit ($)
1,200	0.3	0.08	−0.02	−24,000	−7,200
1,000	0.4	0.05	+0.01	10,000	4,000
800	0.3	0.04	+0.02	16,000	4,800
Total					**1,600**

Box A.3
Valuation of an oil field

Consider the problem of valuing a 1M barrel oil field that we could acquire in 6 months. We know its extraction will cost $75/bbl, and the average estimate of the price in 6 months is $80/bbl. However, the price is equally likely to remain at $80/bbl, drop to $70/bbl, or rise to $90/bbl.

If we focus on the average future price, the value of the field is $5 M = 1M (80–75). What is the value of the field if we recognize the price uncertainty? An instinctive reaction is that the value must be lower because the project is more risky. However, when you do the calculation scenario by scenario, you find that intuition to be entirely wrong!

If the price is $10/bbl higher, the value increases by $10M to $15M. However, if the price is $10/bbl lower, the value does not drop by $10M to –$5M as implied by the loss of $5/bbl when production costs exceed the market price. This is because management has the *flexibility* not to pump and to avoid the loss. Thus, the value of the field is actually 0 when the price is low. The net result is that the actual value of the field would be $6.67M, higher than the $5M estimate based on the average oil price of $80/bbl.

The field is worth more on average, not less. This manifestation of the Flaw of Averages illustrates why it is worthwhile to consider flexibility in design. If management is not contractually committed to pumping, it can avoid the downside if the low oil price occurs while still maintaining the upsides. This flexibility increases the average value of the project compared with an inflexible alternative.

Appendix B: Discounted Cash Flow Analysis

This text refers throughout to discounted cash flow (DCF) analysis, the most common methodology for an economic appraisal and comparison of alternative system designs. Although widely used, DCF has significant limitations in dealing uncertainty and flexibility. In response, some academics call for its wholesale replacement by a different and "better" method.[1] This is not our approach. We wish to build on the widespread use of the DCF methodology. We thus advocate a pragmatic incremental improvement of the method to alleviate its limitations in dealing with uncertainty and flexibility.

To improve DCF sensibly, it is important to understand its elements and procedures. This appreciation supports the use of Monte Carlo simulation to deal with uncertainty, as appendix D indicates.

The purpose of this appendix is to remind readers of the basic principles of DCF:

• Its main assumptions and associated limitations,

• The mechanics of discounted cash flows,

• The calculation of a net present value and an internal rate of return, and importantly,

• The rationale for the choice of a suitable discount rate.

The Issue

Every system requires cash inflow (investments and expenses) and generates cash outflow (revenues) over time. These cash flows are the basis for the economic valuations of system designs. Indeed, economists tend to regard projects or system designs merely as a series of cash flows. They are oblivious to the engineering behind their generation. The question is: How should we value these revenues and expenses over time?

The underlying economic principle is that money (or, more generally, assets) has value over time. If we have it now, we can invest it productively and obtain more in the future. Conversely, money obtained in the future has less value than the same amount today. Algebraically:

X money now → (1+d)*X = Y at a future date

Y at a future date → Y/(1+d) = X money now,

where d > 0 is the rate of return per dollar we could achieve if we invested money over the respective period. The rate of return captures the time value of money. This means that cash flows Y in the future should have less value, that is, be "discounted," when compared with investments X now. This is the rationale behind DCF analysis.

The most fundamental assumption behind a DCF analysis is that it is possible to project the stream of net cash flow, inflow minus outflow, with a degree of confidence over the lifetime of a project or system. To facilitate the analysis, it is usual to aggregate cash flows temporally, typically on an annual basis. Table B.1 shows illustrative cash flows of two system designs.

Which design would you prefer? Design A requires a lower initial investment but annual cash investments of $100M for 3 more years before it is completed and sold off for $625M. Design B requires a substantially larger initial investment but delivers positive cash flows from year 1 onwards. However, design B has to be decommissioned in year 5 at a cost of $310M. Note that if you do not discount income and expenses, design A is a winner (it nets $155M = $625M − $470M), and design B is a loser (it shows a net loss of −$20M = $550M − $570M). This is the perspective of tax authorities, but it does not constitute a proper economic analysis because it does not account for the time value of money.

Table B.1
Cash flow profiles of two designs ($ millions)

Cash flow projections						
Year	Upfront investment	1	2	3	4	5
Design A	−$170.0	−$100.0	−$100.0	−$100.0	$625.0	
Design B	−$260.0	$200.0	$200.0	$100.0	$50.0	−$310.0

A proper economic valuation has to answer two questions:

- Is a particular design worth implementing? Does it have positive economic value?
- If there are several design alternatives, which design is preferable?

The answers to these questions depend on the time value of money, specifically on the availability and cost of the capital needed to finance a project. This is the return the organization will have to pay to its financiers: to its banks in the form of interests on loans and to its owners in the form of dividends. DCF analysis offers a way to answer these questions.

DCF Principle

DCF is based on a rather simplistic view of an organization's finances: All its capital sits in a single virtual bank account. The group finances all its investments from this account. It receives any surplus that the project generates and pays out any shortfall or investment needed. In reality, firms finance their investments from a variety of sources with different conditions attached. They also place capital in different investments with different risks and returns. However, the assumption of a single bank account simplifies comparisons between projects considerably and, importantly, makes these comparisons conceptually independent of the specific financing arrangements. For government agencies, the sources of income and expenses are much more complicated, but the DCF principle applies equally to them and to government projects. In practice, the main difference between government and corporate uses of DCF analysis lies in the choice of discount rate.

The virtual bank account has an interest rate attached, like any bank account. A second simplifying assumption of a traditional DCF analysis is that this interest rate is fixed and the same for both deposits (positive cash flowing back from projects) and loans (cash injections required to run the projects). This is obviously different from real banks. The interest rate of the company's virtual bank account is called its *discount rate*. We discuss this discount rate in more detail later.

Once we accept the financial view of the company or government agency as a virtual bank account, it is natural to value the cash flow stream that a project generates over time as its contribution to the bank account. The analysis considers that the bank account finances any short-

Box B.1
Assumptions of DCF analysis

1. We can confidently estimate annual net cash flows of projects over the project lifetime.

2. Cash flow shortfalls in any one year come from a "virtual bank account," net positive cash flow will be deposited in this account.

3. The project sponsor, a company or government agency, can borrow arbitrary amounts from the virtual account.

4. The interest rate for borrowing is fixed and the same as for deposits. This interest rate is called the discount rate.

Projects and design alternatives should be compared by their net contribution to the virtual bank account over their lifetime.

falls (i.e., negative cash flows), and it receives and earns interest on any positive cash flows, all at a fixed interest rate. Any project is worthwhile if it generates a net positive return. Design A is preferable to design B if A's net return is larger than B's.

DCF Mechanics

One way to quantify the value of a project is to calculate its net contribution to the company bank account at the end of its life. Table B.2 shows this calculation in detail for the two designs in table B.1. (We show the details to explain the process. In practice, analysts use standard spreadsheet functions to get the results without the detail.) The end-of-life contribution of design A is $12 million after 4 years, and the corresponding contribution of design B is $6.3 million after 5 years. Note, from the perspective that recognizes interest, that design B appears worthwhile—as it does not if we fail to take into account the time value of money.

If two projects have the same duration, then their end-of-life contribution is a sensible way of comparing their economic value. However, we generally have to compare projects with differing durations, as in the case of designs A and B in table B.2. In this situation, it is not fair to compare the end-of-life results (e.g., $12 million for design A vs. $6.3 million for design B) because the same amounts in different years are not equivalent. To make the comparisons fair, to compare apples with apples as it were, it is customary to relate the end-of-life contributions of projects to a sum of money at a common time. This common time is

Table B.2
Lifetime contribution of two designs ($ millions)

Lifetime contribution	Discount rate	10%				
Year	Upfront investment	1	2	3	4	5
Design A cash flow	–$170.0	–$100.0	–$100.0	–$100.0	$625.0	
Starting balance		–$170.0	–$287.0	–$415.7	–$557.3	
Interest		–$17.0	–$28.7	–$41.6	–$55.7	
Inflow/outflow of capital at year end		–$100.0	–$100.0	–$100.0	$625.0	
Year-end balance		–$287.0	–$415.7	–$557.3	$12.0	
Design B cash flow	–$260.0	$200.0	$200.0	$100.0	$50.0	–$310.0
Starting balance		–$260.0	–$86.0	$105.4	$215.9	$287.5
Interest		–$26.0	–$8.6	$10.5	$21.6	$28.8
Inflow/outflow of capital at year end		$200.0	$200.0	$100.0	$50.0	–$310.0
Year-end balance		–$86.0	$105.4	$215.9	$287.5	$6.3

normally the present. We thus refer to the *present value* of projects. More specifically, to indicate that we account for the difference between the revenues and expenses, we focus on the *net present value* (NPV).

It is natural to think of today's equivalent of the positive end-of-life contribution of a project as the sum of money we can borrow from the bank account today against the net contribution of the project, its NPV. In other words, the NPV is the amount we can withdraw from the company bank account today (and spend in other ways) so that the end-of-life contribution of the project will allow us to pay back the accrued debt of our withdrawal at the end of the project.

An amount X borrowed from the company account at annual interest rate r will accrue to $X*(1+r)^T$ at the end of T years. This sum has to be equated to the project's end-of-life contribution after year T. So, NPV = (end-of-life contribution)/$(1+r)^T$. Using a discount rate of 10 percent (= 0.1), the NPV of design A is therefore $12.0/(1.1)^4$ = $8.2 million. Likewise, the NPV of design B is $6.3/(1.1)^5$ = $3.1 million.

If the end-of-life contribution of a project is negative, then its NPV is the amount we need to *deposit* in the company bank account today to cover the project's shortfall at the end of its lifetime. Financial theory

Table B.3
Net present values (NPV) of two designs ($ millions)

Net present value—Short version	Discount rate	10%				
Year	Upfront investment	1	2	3	4	5
Design A	−$170.0	−$100.0	−$100.0	−$100.0	$625.0	
Effect today (DCF)	−$170.0	−$90.9	−$82.6	−$75.1	$426.9	
Sum of discounted cash flows (NPV)	$8.2					
Design B	−$260.0	$200.0	$200.0	$100.0	$50.0	−$310.0
Effect today (DCF)	−$260.0	$181.8	$165.3	$75.1	$34.2	−$192.5
Sum of discounted cash flows (NPV)	$3.1					

and common sense suggest that companies should not invest in projects with a negative NPV unless there are significant externalities that offer value beyond the project cash flow, such as access to the client's future more profitable projects.

Short-Cut Calculation

It is common practice to calculate the NPV using a short cut based on discounted annual cash flows. We do this by calculating for each annual cash flow the sum that could be borrowed from or needs to be deposited in the account today to balance out the cash flow in the respective year. These sums are of course smaller than the actual cash flow, which is why they are called "discounted cash flows." The NPV for the entire project is then the sum of these discounted annual cash flows.

For example, design A's cash flow in year 3 is −$100 million. If X is the amount deposited today to balance out this shortfall, then $X*1.1^3$ is the amount available after 3 years. Therefore, the amount X today that balances out the negative cash flow of −$100 million in year 3 is $-\$100/1.1^3$ = −$75.1 million. Table B.3 illustrates the short form of the NPV mechanics for the designs in table B.1.

Choosing the Discount Rate

The discount rate is at the heart of the DCF principle of a virtual bank account. Which discount rate should one choose? As a practical matter, most system designers have to use the discount rate set by higher author-ity. In most companies, the board of directors or the chief finance officer sets a discount rate employees should apply to all projects across the company. The process is similar for government agencies: Somebody is responsible for establishing the applicable rate.[2] Designers may thus not have a choice about the discount rate. However, it is still useful to under-stand the rationale for its selection.

The discount rate should capture the firm's cost of capital: The firm should be able to raise funds at the discount rate and should be able to invest these funds so that the return on these investments equals the discount rate. Suppose a firm can borrow from a bank at 6 percent interest. Does that mean that its discount rate should be 6 percent? The answer is no. In fact, a significant proportion of the firm's capital will come from the owners of the firm rather than from banks. The owners are more vulnerable to the risk of default because debtors have preferential access to liquidation proceeds in the case of bankruptcy. Indeed, the bank will only grant the loan because there are owners who are primarily liable with their invested capital (their "equity") in the case of bankruptcy. The owners therefore take more risk and will demand a higher return from the company than the bank's 6 percent. The cost of the owners' equity, that is, their return expectations, is higher than the cost of debt, which is the interest on loans. To determine a sensible discount rate, we need to derive the average cost of capital by appropriately weighting the cost of equity and debt.

Weighted Average Cost of Capital

The weighted average cost of capital (WACC) provides a reasonable estimate of the discount rate. It represents the average cost return expected by the owners and banks that finance a project. A simple example illustrates the process for its calculation. If the company's total market value (number of shares times share price) amounts to 75 percent of the company's total invested capital and the shareholders expect an annual dividend of 10 percent, and the remaining 25 percent of the total invested capital is financed through debt at an average inter-est rate of 6 percent, then the company's WACC is 75 percent*10 percent + 25 percent*6 percent = 9 percent. In practice, the details of the calcula-tion are more complex and depend on the specific context of the company.[3] Conceptually, the important point to retain is that the proper

Box B.2
NPV rule

> Assumption: The cash flow profiles of projects fully capture their economic value.
>
> A. A project is economical if its NPV is non-negative.
>
> B. If two projects are mutually exclusive, then the project with the higher NPV is preferable.

discount rate is higher than the interest rate paid for loans to finance a project.[4]

Making Decisions with NPV

Recall the fundamental questions: Is a project economical? If projects compete, which one should we implement? The so-called NPV rule stated in box B.2 gives the answers.

A project may be worthwhile even if its NPV is not hugely positive. All the costs of financing the project, that is, the dividend payments and the interests on various loans, are already factored into our setup of the virtual bank account—via the discount rate. Any positive NPV indicates that the project delivers more than the normal, threshold rate of return. If a company continuously produces highly NPV positive projects, the market will realize this and value the company higher, which will lead to higher dividend expectations. This in turn will lead to an adjustment of the discount rate, which will reduce the NPVs of typical projects. A project is desirable if its NPV is positive, even slightly.

Dependence of NPV on Discount Rate

The NPV depends on the discount rate in a rather complex way. Figure B.3 illustrates this by exhibiting the NPVs of designs A and B as the discount rate changes. Design A is much more sensitive to discount rate changes, although it is of shorter duration. Its NPV decreases as the discount rate increases. This makes intuitive sense. Insofar as the discount rate is the company's cost of capital, the more the company has to pay the banks and owners, the less the company keeps. Design B is more stable, but its NPV has an interesting property: its dependence on the discount rate is non-monotonic. For very low discount rates, its NPV is negative. Rising discount rates then lift the NPV above zero (by decreasing the importance of the large closure cost at the end of the project). As discount rates rise further, they adversely affect NPV (by diminishing the value of the positive returns) and it turns negative again for very large discount rates.

Why does NPV depend so much on the discount rate and project life? To answer this question, consider the effects on positive and negative cash flows separately. The contribution of a positive cash flow in year T to the NPV equals the sum of money we can withdraw from the company account today so that the accrued debt in the account in year T will be covered by the positive project cash flow in year T. If the discount rate on the virtual account increases, the hole in the account made by our withdrawal grows faster. The amount we have available in year T, however, will remain the same size. Therefore, we have to make the initial hole smaller, that is, withdraw less money. This deteriorating effect of higher discount rates on the value of positive cash flows is larger the further in the future the cash flow lies. This results from the compounding of interest that we pay on our withdrawal. Figure B.4 shows the effect: Positive cash flow further in the future deteriorates faster as the discount rate increases. In summary, positive cash flows contribute less to the NPV as discount rates increase, and this deterioration is greater for late cash flows.

The effect on negative cash flow is the opposite. As the interest rate on the account increases, the amount we need to deposit today to fill the hole due to the future negative cash flow becomes smaller because our deposit will grow faster. A negative cash flow of $100 in 2 years at a 10 percent discount rate is equivalent to borrowing $100/1.1^2 = \$82.64$ from the bank account today. The same cash flow in 4 years is equivalent to withdrawing $100/1.1^4 = \$68.30$ today. Therefore, as the discount rate increases, it dampens the negative contribution of negative cash flows to the NPV, and this positive effect is greater for late cash flows (see figure B.5). The effects of discount rates on NPV are summarized in box B.4.

In the example in table B.1, design A has a large late positive cash flow, whose contribution is very vulnerable to changing discount rates. This causes the sensitivity of this design's NPV (see figure B.3). However, design B has most of its positive cash flow early on. The negative effect of increased discount rates on these cash flows will be relatively mild

Box B.3
Internal rate of return

A frequently used complement to the NPV method is the calculation of the internal rate of return (IRR). To calculate the IRR, one starts by calculating the net contribution at the end of the project, which depends on the interest rate charged on the virtual bank account, the discount rate. How high an interest rate can be charged before the end-of-life contribution of a project turns negative? Alternatively, equivalently, what interest rate would lead to a zero NPV? This interest rate is called the IRR.

If the discount rate equals the IRR, then the net contribution of the project to the bank account is zero; cash inflows and outflows balance out, accounting for associated interest payments. Accordingly, we should only invest in projects with an internal rate of return at or above their discount rate. If we thus use IRR as a method of appraisal, we often refer to the discount rate as the hurdle rate. The IRR has to exceed the hurdle rate for a design to be worth implementing.

One way of calculating the IRR is to plot the NPV for various discount rates and find the value where the NPV is zero. The result for designs A and B from table B.1 appears in figures B.1, B.2, and B.3.

The IRR of design A is 11 percent, slightly higher than the hurdle rate of 10 percent. Project B is an interesting case. It has two IRRs, one at 7 percent and the other at about 23 percent. In fact, the NPV will be positive for any interest rate between 7 percent and 23 percent. Because the hurdle rate of 10 percent is in that range, the project is economically viable. Multiple IRRs occur only if there is more than one sign change in the cash flow. Put another way, if all negative cash flows or investments happen before the first positive cash flow, then the IRR is unique.

The IRR has to be taken with a pinch of salt. Yes, project B would break even at a discount rate of 23 percent. Does that mean the project makes a 23 percent return? No. It only breaks even with a 23 percent discount rate because it is assumed that we can deposit money today in the virtual bank account at a 23 percent interest rate and then withdraw it in year 5 to cover the substantial decommissioning cost. How likely is that? Recall that the discounted cash flow model assumes that the firm can borrow *and* invest money at the discount rate. If this assumption is unrealistic, then the discounted cash flow model is fundamentally flawed. In that case, the IRR is flawed because it is calculated on the basis of the discounted cash flow model. IRRs, especially when they are high, should not be overrated. They are a useful complement to the NPV but do not replace the more intuitive NPV measure. Specifically, when we rank projects, it is preferable to do so by NPV and not by IRR. We refer readers to the standard finance literature for more details.[5]

Note, finally, that ranking projects by IRR or NPV will not necessarily give the same results!

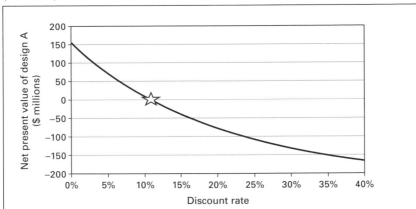

Figure B.1
Net present value for design A as a function of discount rate. IRR is 11 percent.

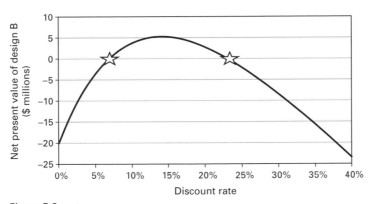

Figure B.2
Net present value for design B as a function of discount rate. IRR is either 7 or 23 percent.

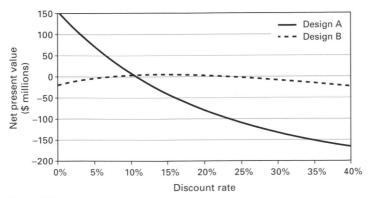

Figure B.3
NPV of two designs for varying discount rates.

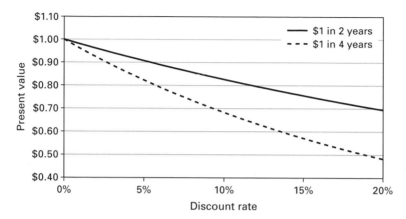

Figure B.4
Present value of $1 occurring in 2 or 4 years for a range of discount rates.

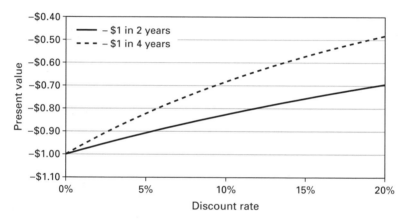

Figure B.5
Dependence of negative cash flows on project life.

Box B.4
Effect of discount rate

> The later a *positive* cash flow occurs and the higher the discount rate, the lower its positive contribution to the overall NPV. The later a *negative* cash flow occurs and the higher the discount rate, the lower its negative contribution to the overall NPV.

because they occur early. At the same time, the project has a large negative cash flow at the end. The increasing discount rate dampens the negative effect of this late payment, so it has an overall positive effect on the NPV. These moderate negative effects on the early positive cash flows and the positive effect on the late negative cash flow balance out and lead to the relative robustness of design B to changes in the discount rate.

Appendix C: Economics of Phasing

Phasing the development of a project is an obvious way to implement a form of flexibility. A design that builds a smaller facility now, instead of a larger one, can give the system owners the flexibility to build additional capacity when needed, at the most desirable size at that time, and in the form and place suitable to later circumstances. Moreover, building up capacity in phases defers the costs of construction and thus saves on interest payments and opportunity costs. The garage case discussed in chapters 1 and 3 illustrates these points.

However, there are also good reasons to build facilities all at once, rather than phasing their development. Obvious reasons to avoid phasing are that the development of a second phase may disrupt ongoing operations of the first phase and require duplicative design and implementation efforts. More importantly, but more subtly, the existence of economies of scale means that it may be cheaper per unit of capacity to build a single large facility rather than two or more smaller facilities.

Overall, the optimal policy for phasing depends mostly on three factors:

• *the time value of money*, which motivates the deferral of costs so as to minimize the present value of costs;

• *economies of scale*, which contrarily provide the incentive to build single big facilities at once, rather than smaller facilities developed in phases; and

• *learning effects* that counterbalance economies of scale to a degree, in that they reduce the cost of implementing second and later editions of the same initial facility.

The optimal solution depends on the specifics of any particular situation, and there is no general solution to the question: What is the best policy for phasing the development of system capacity?

This appendix explains the nature of the three main factors that influence phasing in the simple case when uncertainty is not taken into account. It provides a basis on which we can understand the role of flexibility in design and implementation of projects and products. The idea is to help readers understand how these factors shape and limit the range of design choices.

Time Value of Money

Absent any economies of scale or overhead costs for each transaction, the most economical policy is to create capacity as needed. Don't buy today what you can put off until tomorrow. This is because of the time value of money.

It is cheaper, based on present values, to pay for something later on at the same price than to pay for it now. The present value of a future expense of C is $C/(1 + r)^T$ where r is the discount rate and T is the number of periods. Because both r and T are positive numbers, the present value of a deferred cost is thus less than the immediate value of that cost.

The obvious exceptions to the rule that it is more economical to defer costs is due to high transaction costs and inflation. If the cost of setting up or carrying out a purchase are significant, we may want to reduce the number of purchases. Thus, a person living far from a store may buy food for several days ahead, rather than just for immediate use, to minimize the total expense or bother of going to the store. Also, if the inflation, the rate of price increase, associated with particular commodities is higher than the discount rate, then it may be advantageous to stockpile these goods.

The more important exception to the rule that it is desirable to defer costs is associated with the phenomenon of economies of scale. As detailed below, this is what may lead us to buy the "large economy size" because buying the larger size costs less per unit of capacity.

Economies of Scale Effects

The Concept

The concept of economies of scale is that it is cheaper per unit (the "economies") to acquire capacity in larger sizes (the "scale"). We usually express this in terms of total costs:

Total Cost of System = $K \, (\text{Capacity})^A$,

where A is the economies of scale factor.

We speak of economies of scale when $A \leq 1.0$. The formula then says that the total cost of providing capacity increases less rapidly than the size of the facility. This means that

$$\text{Average cost of capacity} = \text{Total Cost/Capacity} = K \, (\text{Capacity})^{A-1}$$

When $A < 1.0$, the unit cost of capacity keeps decreasing as the plant size gets bigger. Note that the smaller A, the larger the economies of scale. Specifically, increasing capacity by 1 percent increases costs only by A percent.

Economies of scale are pervasive in many industries and technologies. They frequently arise because of the technological reality that costs are roughly proportional to the surface of a facility and capacity proportional to the volume. This is a principal driver of economies of scale in the production of thermal power (using boilers), in chemical industries (more containers and pipes), in transportation (pipelines, ships, aircraft), and in other industries. In short, economies of scale are pervasive in technological industries, and this drives engineering designers toward building large facilities such as large petrochemical and power plants, jumbo aircraft, supertankers, and the like.[1]

The degree of economies of scale is limited in practice. Analyses of real cases rarely indicate economies with A as low as 0.7, and the lower bound on A may be around 0.6. Cases in which A appears to be lower than that are likely to be due to design errors, specifically to failures to adjust system components to different sizes, as illustrated in box 5.6.

Furthermore, we should note that the economies of scale formula overstates the case for developing large systems all at once. The expression assumes that the capacity installed is used. This is generally not true in the initial stages. Indeed, if we build in advance of need, we are deliberately creating capacity that will be idle until the need becomes manifest. When part of the capacity installed is not used, the average cost of the unit of capacity used increases as the cost of the large facility spreads over fewer units. Thus, the actual economies of scale are often considerably less than their potential. A proper analysis of the effect of economies of scale on phasing must account for the fact that capacity built in advance of need is unused for some time. This reality increases the average cost per unit of capacity used and counteracts to some degree the effects of economies of scale.

Optimal Phasing Considering Economies of Scale

Manne (1967) presented an optimal solution for balancing the counter-vailing forces of economies of scale—which promote larger additions to capacity—and the time value of money —which provides the incentive to build smaller. His specific solution is valid only for cases in which growth in demand is steady and constant and there are no transaction costs. His result illustrates the situation rather than provides a general solution. It is useful to consider because it provides insight into how the competing forces balance out.

Figure C.1 shows the crux of Manne's solution. It defines the optimal size of a phase in terms of optimal number of years of growth in capacity it will accommodate. The idea is that each phase of expansion accom-modates capacity growth for some number of years, and this number is the "cycle time." As figure C.1 shows, higher time values of money push down the desirable size for phased additions. Conversely, larger econo-mies of scale (denoted by lower values of the A factor) increase the optimal size of each phase. This graph neatly presents the overall picture for the interaction of economies of scale and the time value of money.

Manne's work indicates that the optimal cycle time is relatively small. If we consider that the time value of money for businesses is typically around or above 10 percent (in real terms, net of inflation), and that the economies of scale might be in the range of 0.7 to 0.8, then the optimal cycle time is only about 6 years. Note that for government agencies that

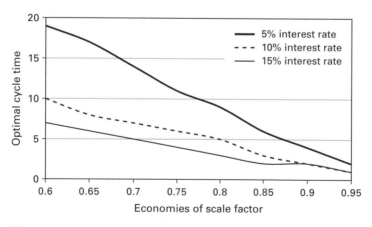

Figure C.1
The optimal cycle time is longer, and thus the optimal initial investment is larger, when the economies of scale are greater and the interest rates r are lower (no learning effect).
Source: Manne, 1967.

use lower discount rates, such as those around 5 percent for in the United States (U.S. Office of Management and Budget), the most economical size of each phase might be about twice as big.

Learning Effects

The Concept

Learning refers to the commonly observed phenomenon in many industries that we manage to produce things at less cost as we produce more of them (see e.g., Wright, 1936; Lieberman, 1984; Terwiesch and Bohn, 2001). The general explanation for this effect is that as we repeat operations, we learn more about how to do things and work more efficiently. In parallel, we may also alter the design to facilitate production, eliminate waste, and so on. We may also introduce new production technology. So, the phenomenon is not uniquely associated with learning. Yet we refer to these several reinforcing tendencies as the learning curve effect.

Learning curve effects can be spectacular. A major oil company recently reported savings of about 20 percent per unit when it found itself forced to build three similar platforms to develop a remote oil field. For long production runs, the cost per unit can drop by half, compared with that of the first unit. This phenomenon provides great incentives to install capacity consisting of many smaller units instead of a few large units. In short, it tends to counteract the effects of economies of scale.

Analyses of the learning curve phenomenon commonly assume that improvements are continual and go on indefinitely. In some industries, this conclusion seems plausible (Argote et al., 1990). However, this is not a proven rule. Nonetheless, system designers may expect that some form of learning does exist as they ramp up production and deploy more units. They thus should keep this phenomenon in mind.

A common way to model learning assumes that costs drop by a fixed percentage each time capacity doubles. This reduction in unit costs per doubling of capacity is the "learning rate." Thus, if the learning rate is 5 percent, the unit cost of building the fourth plant is 95 percent of the unit cost of the second plant. In practice, researchers estimate the actual learning rate for any industry or product by examining the historical record.

The functional form for this learning curve is:

$$U_i = U_1 i^B$$

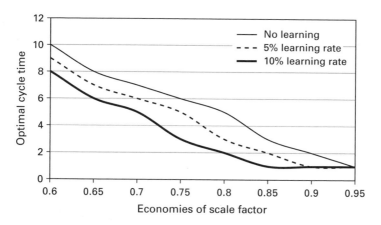

Figure C.2
Higher learning rates reduce the optimal cycle time for phasing of projects (10% interest rate).

where U_i is the production cost of the ith unit, U_1 the cost of the first unit, and B is the slope of the learning curve. We calculate this slope from the empirically observed learning rate, L, as follows:

B = log (100 percent – L percent) / log (2)

Thus, if it costs $100 to produce the first unit, the cost of producing the 32nd unit with a learning rate of 15 percent is:

B = log (100 percent – 15 percent) / log (2) = –0.2345

so that

$U_{32} = \$ 100 (32)^{-0.2345} = \$44.37$

Such reductions in unit costs associated with producing many small increments to a system counteract the advantages of the economies of scale achieved by developing large additions to capacity. Figure C.2 illustrates the results. It is a modified version of figure C.1 that includes learning effects. It shows that the effects of the learning curve counteract those of economies of scale: the larger the learning, the shorter the optimal cycle time. Majd and Pindyck (1989) examine this effect in detail.

Take Away

The analysis of phasing based on the admittedly unrealistic assumption of known steadily increasing demand provides interesting insights into the factors that drive the costs and benefits of phasing. Both the time value of money and the learning curve phenomenon provide the incentive to defer expenses and to phase investments in system capacity. The existence of economies of scale pushes the other way, motivating the construction of larger units of production. The optimal policy concerning the possible phasing of system capacity depends on the relative strength of these factors, as well as important related issues, such as the overhead costs of deploying separate phases.

When demand is uncertain, phasing offers the added advantage of maintaining flexibility in timing and sizing of future capacity additions depending on how demand evolves.

Appendix D: Monte Carlo Simulation

Converting a Standard Valuation into a Monte Carlo Simulation with Flexibility

Chapter 3 introduces Monte Carlo simulation as a preferred methodology for the study of the value of flexibility. Part II refers to this methodology throughout. Monte Carlo simulation models efficiently generate thousands of futures, run all these futures simultaneously through a model of system performance, and summarize the distribution of possible performance consequences graphically.

The purpose of this appendix is to provide a roadmap to the conversion of traditional static system performance models into Monte Carlo models. The aim is to clarify these models, which can be daunting for designers, engineers, or clients who are not familiar with the technique, and who rely on simpler static models such as standard NPV spreadsheets. Along the way, we introduce a variety of techniques to model uncertainty in spreadsheets.

Specifically, this appendix takes you through the conceptual steps that lead from a static valuation model to a Monte Carlo model that we can use to articulate the value of flexibility:

- *Step 1* Produce a standard valuation model
- *Step 2* Perform a standard sensitivity analysis: Change one variable at a time
- *Step 3* Perform a probabilistic sensitivity analysis: Change all variables simultaneously
- *Step 4* Introduce distributional shapes for uncertain numbers
- *Step 5* Introduce dependence between uncertain numbers

- *Step 6*   Introduce dynamically changing uncertain numbers
- *Step 7*   Model flexibility via rules for exercising flexibility

To illustrate these steps, we use the parking garage illustration of chapter 3 as a working example.

A Note on Software

As in the rest of the book, the focus is on spreadsheet models because of their ubiquity. The appendix assumes the reader is familiar with standard formulas and commands in Microsoft Excel®. We can perform Monte Carlo simulations in that program. Without additional software however, Monte Carlo modeling in Excel can become cumbersome, and simulations of larger models tend to take a long time. Commercial software packages facilitate the modeling effort and speed up execution. Examples of commercial packages include @Risk®, Crystal Ball®, XLSim®, and RiskSolver®.

All Monte Carlo software packages have three main components:

• A collection of built-in random number generators, which are special spreadsheet formulas that allow sampling from prespecified distributions. For example, the formula "= gen_normal(0,1)" in XLSim draws a sample from a normal distribution with mean 0 and standard deviation 1 and puts it in the cell that contains the formula. Every time the spreadsheet is recalculated (e.g., when a cell has been modified and the enter key is hit, or when the recalculate key, the F9 function key in Excel, is hit), another number is drawn from this distribution and appears in the cell.

• A convenient and fast way of executing thousands of scenarios and storing the results.

• An interface that facilitates the calculation of summary statistics and the generation of charts to analyze, visualize, and communicate the results of simulations.

Unfortunately, different software packages use different formulas for their random number generators, and their spreadsheet models are not always portable. Because we do not wish to limit our readers to a specific package, this appendix uses standard Excel commands to generate distributions for uncertain numbers. These commands are slower and more cumbersome than the specialized formulas of commercial packages, but the resulting spreadsheets work with all packages. Readers who use

Monte Carlo simulations frequently should invest in a commercial package. For convenience, we used the XLSim® software package to produce some of the graphs in this appendix.[1]

Step 1: Produce a Standard Valuation Spreadsheet

Valuation models for system designs are input–output models. They capture how a system design:

- Converts system inputs, such as capital, labor, material, energy, demand,
- Within constraints (e.g., of a physical, economic, regulatory, or legal nature),
- Into outputs.

Some system outputs will be desirable, such as profit, demand satisfaction, and better health, whereas others will be undesirable, such as congestion or pollution. Roughly speaking, the inputs and constraints describe the world the system faces, whereas the outputs describe the difference that the system makes. The system itself is described by (i) a set of design parameters, such as capacity, productivity, reliability, etc.; and (ii) a set of formulas that relate inputs and system parameters to outputs. Figure D.1 illustrates a valuation spreadsheet for the parking garage example.

Modeling Tips

A few tips on spreadsheet modeling are in order at this point. The advantage of a spreadsheet is that we can hard-wire calculation steps into it via formulas. This is useful when valuation assumptions change. Suppose you have calculated an NPV for a complex project using a 10 percent discount rate when the finance director tells you that the plan needs to be recalculated using a 12 percent rate. If you were using an electronic calculator to compute the NPV, you would have to perform the same calculations all over again using this 12 percent rate. If you use a well-programmed spreadsheet, you simply change the number in the cell that contains the discount rate from 10 percent to 12 percent, and the spreadsheet performs the update instantly.

To exploit this advantage of the spreadsheet, you have to dedicate a single cell to contain the discount rate and refer to it whenever a calculation uses the discount rate. If, however, you place the discount rate, say

Input Table

DEMAND PROJECTION

Demand in year 1	750	spaces
Additional demand by year 10	750	spaces
Additional demand after year 10	250	spaces

REVENUE

Average annual revenue	$10,000	per space used

COST

Average operating costs	$3,000	per space available
Land lease and other fixed costs	$3,330,000	p.a.
Capacity cost	$17,000	per space
	10%	growth per level above level 2
DISCOUNT RATE	10%	

System Parameters

Capacity per level	200	cars
Number of levels	6	levels [DESIGN PARAMETERS]

Performance Calculation

Year	0	1	2	...	15
Demand		750	893	...	1,634
Capacity		1,200	1,200	...	1,200
Revenue ($M)		$7.5	$8.9	...	$12.0
Operating costs ($M)	$0.0	$3.6	$3.6	...	$ 3.6
Land leasing and fixed costs ($M)	$3.3	$3.3	$3.3	...	$ 3.3
Cash flow ($M)	–$3.3	$0.6	$2.0	...	$ 5.1
Discounted cash flow ($M)	–$3.3	$0.5	$1.7	...	$ 1.2

Present value of cash flow ($M)	$26.7
Capacity cost up to two levels ($M)	$ 6.8
Capacity costs levels above 2 ($M)	$17.4
Net present value ($M)	$ 2.5

Figure D.1
Inputs, system parameters, and performance calculation (NPV) for parking garage with six levels.

Box D.1
Modeling vocabulary

Input Numbers that describe the environment that the system will face

System parameters Numbers that describe the system design

Outputs Numbers that describe the performance of the system

---

of 10 percent, in cell A1, you should always refer to A1 and never use the numerical value of the discount rate. To discount the value in cell A2 by 10 percent, you should write "=A2/(1 + A1)." If you use "=A2/1.1" instead, your spreadsheet will not update correctly when you increase A1 from 10 percent to 12 percent. If you have calculated your NPV in a spreadsheet with the numerical discount rate mixed into formulas, as in "A2/1.1," you will have to find all the cells that contain this rate and change them manually—a process that may well take more time than repeating the calculation with an electronic calculator. You have given away the main advantage of the spreadsheet—an instantaneous update when inputs change.

The most important rule in spreadsheet modeling is therefore that every numerical input gets one and only one dedicated cell. Then whenever you need to use that input in a calculation, you reference that respective cell. Never mix formulas and numbers. It is useful to dedicate a range of the spreadsheet, or indeed a separate worksheet, as the input range that contains all the numerical values. The remaining space is the model range; all its cells contain formulas or cell references only. If the layout of the spreadsheet requires inputs at various places, then an alternative to a dedicated input range is suitable color-coding of all cells that contain a numerical value.

A second important point in spreadsheet modeling is that it is easy to make mistakes, to commit "slip of the keyboard" errors. Therefore, models need to be carefully validated. Read every formula several times. If you use a complex formula, input the same formula again into an adjacent cell to verify that it produces the same result. Once you include a formula, check that it behaves as you expect (e.g., see that it does not produce negative values when it should not). Do this by changing the inputs the formula uses to values that may be unrealistic but for which you know what the formula result should be (e.g., the revenue for zero demand should be zero). Validate as you build the model and then again when the model is finished. Change the inputs to see whether the outputs behave in the expected way. Finally, if you have the resources, the best way to validate a model is to ask a second modeler to produce an independent model of the system. Although this will not avoid systematic mistakes due to a shared misunderstanding of the system behavior, it will greatly reduce the chance of a slip of the keyboard error.

A third important point is to document the model well. Use text boxes and cell comment boxes. Use more of these boxes rather than less. Regard the model as a written piece and make it easy for a "reader" to follow the line of reasoning in the model.

Step 2: Perform a Standard Sensitivity Analysis: Change One Variable at a Time

Every model uses assumptions that can be called into question. The parking garage model invites questions such as, "What if demand is lower than expected?," "What if the average annual revenue per space used is lower than projected because there is more use of discounted long-time parking?," "What if operating costs exceed projections?," and "What if the maximal capacity utilization due to variation in demand is lower than anticipated?"

A first step toward the recognition of uncertainty is to understand the effect of deviations of individual inputs from their baseline assumptions. This process is called sensitivity analysis and should be a routine part of practical valuation procedures. Static models as developed in step 1 require you to make fixed base projections for unknown inputs, and sensitivity analyses require you to specify ranges around these base projections. To carry out this analysis, you keep all inputs except one at their base values and alter the free input to track corresponding changes of performance measures as its value changes over its range. The data table command performs this analysis efficiently. In fact, it is arguably the most useful command in Excel. Everyone who works with valuation spreadsheets should familiarize themselves with data tables.

Figure D.2 shows a specific sensitivity chart for the parking garage case. The original model assumed an initial demand of 750 spaces, an additional demand by year 10 of 750 spaces, and then a further additional demand beyond year 10 of 250 spaces. We used the data table command to analyze the effect of deviations from these assumptions by ±50 percent.

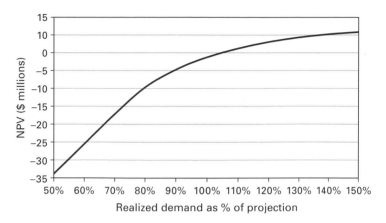

Figure D.2
Sensitivity chart for parking garage (base case = 100 percent).

Note that the graph shows an important asymmetry. Low demand has a more pronounced effect on the NPV than high demand. This is a consequence of the fact that we cannot capture high demand when the garage is at full capacity. As chapter 3 and appendix A explain, such asymmetries cause the Flaw of Averages. That is, the NPV based on base-case conditions is not the average of the NPVs obtained as the condition varies around the base value. Therefore, whenever you generate sensitivity graphs that are not straight lines, this is a sign that there is Flaw of Averages in the system.

Such intuitive interpretations of sensitivity graphs are important validation checks. If you cannot make sense of the shape of a sensitivity graph, it is quite likely that there is a bug in your spreadsheet. You should also be suspicious when you generate straight-line sensitivity graphs and check that you have not missed a constraint in the system.

Tornado Diagram

The tornado diagram shown in figure D.3 is a useful tool for sensitivity analysis. This graph summarizes the relative effects of variations of several inputs over their ranges under the assumption that the other variables remain at their base values. Several freeware spreadsheet add-ins are available to facilitate the production of tornado diagrams.[2]

The tornado diagram illustrates which uncertain inputs most affect the relevant performance metric. Each variable has an associated bar representing its impact on the performance metric as it varies over its

Input	Base Case	% Sensitivity	Low	High
Realized demand (as % of projection)	100%	50%	50%	150%
Average annual revenue	$10,000	20%	$8,000	$12,000
Average operating costs	$3,000	15%	$2,550	$3,450
Capacity cost	$13,600	10%	$12,240	$14,960
Average capacity utilization	80%	10%	72%	88%
Land lease and other fixed costs	$3,330,000	5%	$3,163,500	$3,496,500
Growth of capacity cost above level 2	10%	20%	8%	12%

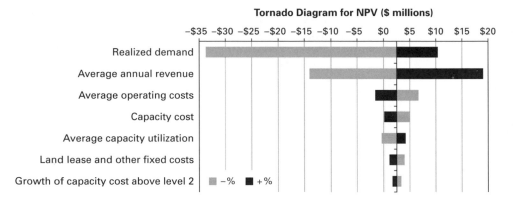

Figure D.3
Input ranges and associated NPV tornado diagram for parking garage.

prescribed range. The diagram sorts the bars from top to bottom by their length, going from longer ones at the top to shorter ones at the bottom, so that the result looks like a funnel—hence the name tornado diagram. The graph is particularly useful if there is a large number of uncertain inputs in the valuation model and it is impractical to analyze them all carefully. The tornado diagram provides a way to prioritize and choose the most important uncertainties to consider: the variables associated with the longest bars.

Tornado diagrams have a second important advantage. Just as sensitivity graphs, they allow the detection of asymmetries and therefore potential Flaws of Averages. In our example, the tornado diagram in figure D.3 shows that when demand varies symmetrically by ±50 percent, its effect is skewed: The bar to the right is shorter than the bar to the left of the base value. Equal changes in this input lead to unequal changes in performance, which is the indication that the Flaw of Averages is at work.

Box D.2
Sensitivity analysis vocabulary

> *Sensitivity analysis*   Replaces fixed assumptions on inputs by ranges on inputs and produces graphs that show how system performance changes as individual inputs change over their range, with all other inputs fixed at their base case values
>
> *Tornado Diagram*   A bar chart that summarizes the effects of the changes of variables across specified ranges

Step 3: Perform a Probabilistic Sensitivity Analysis: Change All Variables Simultaneously

A considerable drawback of traditional sensitivity analysis is that it inspects the effect of changes only one input at a time, holding the other inputs constant.[3] In mathematical terms, standard sensitivity analysis is akin to estimating the partial derivatives of performance measures with respect to the input variables. This variable-by-variable information only provides a good approximation of performance sensitivity close to the base values of the inputs.

Simultaneous effects when two or more variables differ jointly from their projected values can be very important, in particular when their ranges are large. For example, consider net revenues as the product of margin and sales volume. If the projected margin is very small, then additional sales do not greatly increase net revenues. In the extreme case of a zero margin, additional sales have no impact on net revenues. So traditional sensitivity analysis may conclude that sales volume uncertainty has little effect on performance when the base-case margin is low. However, if margin and sales volumes grow simultaneously, they reinforce each other, which can have a significant effect—one that standard what-if analysis would overlook.

Probabilistic sensitivity analysis alleviates the problem of one-dimensional sensitivity analysis. This technique sits between standard sensitivity analysis and a full Monte Carlo simulation. Just as standard sensitivity analysis, probabilistic sensitivity analysis works only with ranges. It makes no specific assumption on the relative likelihood of these values (i.e., all values over a specified input range are deemed equally likely). However, in contrast to standard sensitivity analysis, it simultaneously randomly samples inputs from their respective ranges.

A particular trial of a probabilistic sensitivity analysis therefore consists of a random choice for each input from its range. The process records the inputs and their associated outputs and repeats the sampling many times. The advantage of probabilistic sensitivity analysis is that it explores the effect of joint changes in the inputs. It is thus more realistic than the traditional one-at-a-time sensitivity analysis.

Creating the Probabilistic Analysis

To build a model for probabilistic sensitivity analysis, it is necessary to develop a way to sample the ranges of variation. To do this we can use Excel's random number generator, the RAND() function. A cell that contains "=RAND()" will contain a number between 0 and 1. Moreover, this number updates when the spreadsheet recalculates. A Microsoft Excel spreadsheet recalculates every time a cell is changed. It also recalculates when the function key F9 is hit. Hitting F9 repeatedly is like rolling dice for the cell with the "=RAND()" formula. If you recalculate the spreadsheet many times, say several thousand, you will find that the numbers in the cell spread uniformly over the interval 0 to 1.

The RAND() function can be used to sample an uncertain input from its range. Suppose the lower bound of an input is in cell A1 and its upper bound in cell B1. If cell C1 contains the command "=A1+RAND()* (B1-A1)," it will contain a number between A1 and B1. Every time the function key F9 is hit, a different number will occur. The RAND() function ensures that the numbers between A1 and B1 have, for all practical purposes, the same chance of being sampled into cell C1.

When model inputs change, the performance of the system changes as well. Monte Carlo simulation refers to a computer program, or spreadsheet, that allows automatic sampling and recording of input values and associated output values. Figure D.4 shows the first 20 trials of a Monte Carlo simulation for the parking garage example. Each row refers to one hit of the F9 key (i.e., one combination of inputs, sampled randomly from their prescribed ranges). The results are recorded in columns with sampled inputs first, followed by generated outputs (NPV in this case). With Monte Carlo software add-ins, it takes literally seconds to generate these data. An alternative, using standard Excel, is to employ the data table command.[4]

Correlation Analysis

Once we have generated the Monte Carlo data, we can analyze them statistically to provide additional understanding of the sensitivity of the

Trial number	Demand deviation	Average annual revenue	Average operating costs	Land lease and other fixed costs	Capacity cost	Growth of cap cost above 2 levels	Average capacity utilization	NPV
1	77%	$10,090	$2,702	$3,356,802	$13,546	12%	77%	−$6,200,704
2	101%	$10,012	$2,899	$3,476,480	$12,370	11%	79%	$3,812,784
3	120%	$8,415	$3,322	$3,405,712	$13,391	9%	78%	−$10,122,984
4	70%	$9,280	$3,178	$3,416,020	$14,945	11%	85%	−$26,960,275
5	81%	$11,100	$3,092	$3,402,570	$14,570	11%	82%	−$2,226,732
6	149%	$10,045	$2,624	$3,227,895	$13,278	8%	82%	$17,619,282
7	122%	$8,672	$3,369	$3,474,922	$12,871	9%	87%	−$4,807,860
8	109%	$11,305	$2,737	$3,264,693	$12,447	10%	88%	$24,872,971
9	136%	$9,395	$3,267	$3,258,282	$13,380	10%	86%	$5,461,836
10	105%	$10,391	$3,148	$3,358,847	$14,670	10%	77%	$2,611,193
11	57%	$11,097	$3,169	$3,489,858	$14,343	9%	76%	−$24,226,210
12	107%	$8,805	$2,982	$3,459,562	$12,272	10%	77%	−$5,233,220
13	56%	$10,418	$3,117	$3,221,068	$13,715	11%	87%	−$28,868,329
14	110%	$10,049	$2,924	$3,394,503	$12,242	8%	81%	$9,103,321
15	76%	$8,928	$2,800	$3,251,731	$12,701	10%	82%	−$13,775,546
16	77%	$8,346	$2,853	$3,175,547	$14,407	11%	75%	−$19,388,662
17	61%	$11,986	$3,112	$3,421,734	$13,964	9%	88%	−$16,791,780
18	137%	$8,099	$3,084	$3,454,519	$14,751	11%	79%	−$12,350,919
19	148%	$11,571	$2,643	$3,246,171	$13,305	11%	84%	$31,867,573
20	101%	$11,435	$2,778	$3,235,438	$12,378	10%	87%	$22,473,298

Figure D.4
Data table with first 20 trials of a simulation run for the parking garage example (ranges as in figure D.3).

Box D.3
Monte Carlo simulation vocabulary

Uncertain inputs   The cells in the spreadsheet whose uncertainty significantly affects system performance.

Input distribution   Distribution of the uncertain inputs, including their relationships and dynamic evolution. The determination of a suitable distribution for the uncertain inputs is the result of a dynamic forecasting exercise, as appendix E explains.

Output distribution   Distribution of performance metrics as a function of the input distribution.

Trial   A single run of a Monte Carlo simulation model, sampling one input combination from the defined distribution and recording the associated values of all relevant output cells, calculated by the valuation model.

Monte Carlo simulation   A list of many sampled input combinations and associated calculated output metrics, ready for statistical analysis and graphical display.

relationships between inputs and outputs. Correlation analysis is a way of studying the dependence of outputs on the various inputs. As the tornado diagram, it helps prioritize uncertain inputs, identifying those that contribute most to the uncertainty of the NPV. Figure D.5 shows the correlation between the NPV and the uncertain inputs for the garage case, and it confirms that demand uncertainty is the input that affects its performance most significantly.

Correlation analysis has the advantage of being more realistic than a tornado analysis. It allows all uncertainties to vary simultaneously over their ranges, and not one by one with the others kept fixed at their base values, as in the alternative tornado diagram.

Correlation has the disadvantage of being an unintuitive concept. It measures the degree to which two uncertain numbers are linearly related. However, uncertain numbers with a low correlation can still be closely related nonlinearly. This is an issue because nonlinearities between system inputs and outputs occur quite often, for example, because of system constraints. Take for example an inventory of 2,000 parts. Suppose demand ranges between 1,000 and 3,000 parts with all values equally likely. If demand is lower than 2,000, then there is waste, which costs $1,000 per part; if demand is higher than 2,000, then there are lost sales, which are again costly, say also $1,000 per part. Therefore, cost = |demand − 2,000 units|*$1000/unit. Because of the absolute value function |.| the output (cost) is a nonlinear function of the input (demand). There is a clear relationship between demand and cost, as figure D.6 shows, yet the correlation coefficient of the generated pairs of demands and costs is about 0.03, statistically indistinguishable from zero.[5]

Scatter Plot

A scatter plot matrix is a better tool for the analysis of relationships between input and output variables. It simply replaces each correlation

Correlation coefficients	Demand deviation	Average annual revenue	Average operating costs	Land lease and other fixed costs	Capacity cost	Growth of cap cost above 2 levels	Average capacity utilization
NPV	0.64	0.49	−0.17	−0.01	−0.10	−0.03	0.03

Figure D.5
Correlation between NPV and uncertain inputs for the parking garage example (ranges as in figure D.3).

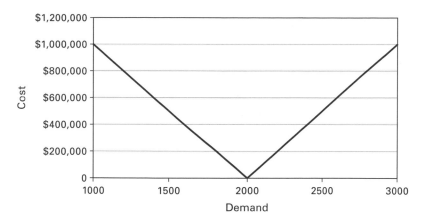

Figure D.6
A nonlinear relationship between inventory cost and demand with a vanishing correlation.

coefficient in a correlation matrix by a scatter plot of the two relevant variables.[6] Scatter plots are more informative and more intuitive than correlation coefficients.

Figure D.7 shows two scatter plots generated from Monte Carlo output for the parking garage, with *all* inputs sampled simultaneously from the ranges given in figure D.3. Notice that the range of the performance metric on the vertical axis, the NPV, is the same for both plots, whereas the horizontal axis covers the range specified for the input variable. It is important to make the vertical axis the same range to facilitate comparisons between such scatter plots.

The first observation is that neither of these two inputs has a dominating effect on the overall uncertainty. The residual NPV uncertainty, driven by the remaining input uncertainties and depicted by the vertical spread of the points cloud, is substantial for both scatter plots. Nevertheless, we recognize that demand uncertainty has a more pronounced effect. The uncertainty in the remaining inputs almost entirely wipes out the effect of the uncertainty in capacity utilization.

Figure D.7 shows that the relationship between demand and NPV is nonlinear, roughly following the curve in figure D.2. As indicated earlier, this is because capacity constraints cut off the benefits of high demands. The scatter plot also shows that the vertical spread of the points to the right is larger than to the left, that is, the expected NPV as well as its residual uncertainty, driven by the remaining uncertain inputs, increase when demands are larger than expected.

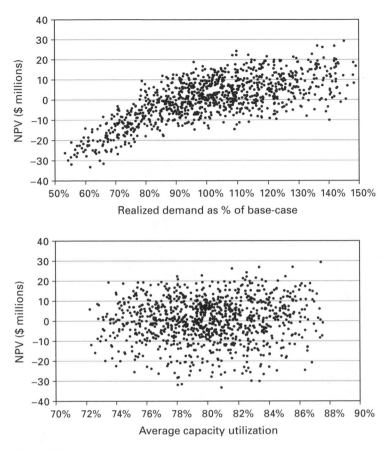

Figure D.7
Scatter plots of NPV versus uncertain input for 1,000 Monte Carlo trials. The vertical variation parallel to the y-axis is due to variability of the other uncertain inputs.

Optimization in the Context of Uncertainty

Optimization, finding the "best" set of system parameters, is an important step in the design process. In our illustrative garage case, the number of levels is the only design parameter. Going back to our fixed projection model in figure D.1, we can find the optimum by varying the number of levels, as in figure D.8. On this basis, it seems sensible to build six levels, albeit the difference to the five-level garage is not large.

An alternative optimization technique is to simply sample the design parameters, here the number of levels, together with the other uncertain inputs from sensible intervals, say between four levels and seven levels

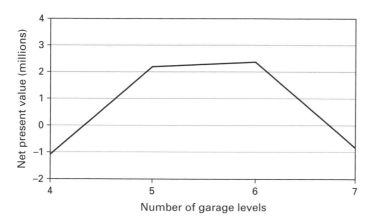

Figure D.8
Optimizing the number of levels of the parking garage.

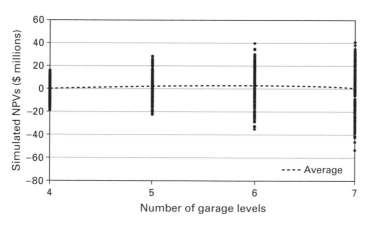

Figure D.9
Optimization under uncertainty. Each point corresponds to a different simulated future. Notice how figures D.8 and D.9 differ in scale.

in this case. The "RANDbetween" function in Microsoft Excel allows you to sample *integers* between any two given integers, giving each the same probability. "=RANDbetween(1,8)," for example, gives every integer between 1 and 8 the same probability (1/8) of being selected. Figure D.9 shows the result for the parking garage. It makes clear that the choice among levels 4, 5, 6, or 7 could be dominated by the uncertainties in outcome, especially because the possible losses are 10 times the expected NPVs.

In general, designs "optimized" for deterministic cases are often sub-optimal when we recognize uncertainties and understand their effects. A second example reinforces this message. This case comes from a service industry and concerns the optimization of staffing levels, which can vary continuously (see figure D.10). Whereas the static base case NPV model seemingly provides clear optimization guidance about the optimum staffing level, the more realistic probabilistic model shows that there is no point in thinking too much about precise optimization relative to the uncertainty in the environment. The base-case NPV model gives the false impression that there is a clear-cut optimal solution—and this is not right.

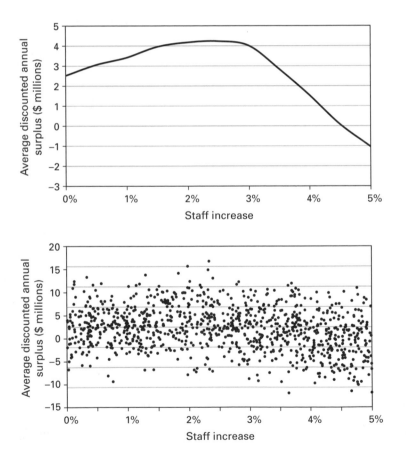

Figure D.10
Example of optimization of staffing level using base-case inputs versus probabilistic inputs. Notice how the two graphs differ in scale.

Step 4: Introduce Distributional Shapes for Uncertain Numbers

Step 3 extends traditional variable-by-variable sensitivity analysis to a simultaneous sensitivity analysis that samples thousands of input combinations from their respective ranges. However, that analysis assumes that all input realizations within a given range are equally likely. This is unrealistic. We may have good reasons to believe that some possible realizations are more likely than others. For example, realizations in the center of the range may be regarded as more likely than those at the extremes. The accommodation of differential likelihoods of inputs over their range is the essence of step 4—in fact, of Monte Carlo simulation.

Histogram

A histogram is the most common means to display differential likelihoods. It is a bar chart of the distribution of uncertain values. To obtain it, we first divide the range over which a variable can vary into a number of regions of equal width, called bins. To each bin we then allocate the fraction of realizations of the uncertain number that we would expect to find in it if we sampled many times. Note that these frequencies must add up to 100 percent, and the bins must be of equal width to have the same a priori chance of capturing the realization of an uncertain number.

Figure D.11 compares the theoretical histogram for the "RAND()" function in Microsoft Excel with an experimental histogram based on 10,000 samples from "RAND()." These histograms are quite although not perfectly similar. They do not exhibit differential likelihoods. They are equal for practical purposes.

The histogram in figure D.12 assigns different likelihoods to different regions of the range over which the uncertainty varies. Values in the middle are more frequent than values at either end. This shape is a triangular distribution. We can easily implement symmetric triangular distributions in Excel as the sum of two uniform distributions.[7] In some cases, it may make sense to use a nonsymmetric triangular distribution, with the peak not in the middle of the range. Generating such distributions with the RAND() function, although possible, is more cumbersome as box D.4 indicates. This is where Monte Carlo software packages add value; they make it easy for users to generate uncertain numbers from many different shapes of distributions.

Software catalogs of histogram shapes provide a versatile tool to express uncertainty in inputs. However, the greater the choice, the more difficult it is to choose among the shapes. In this connection, one should

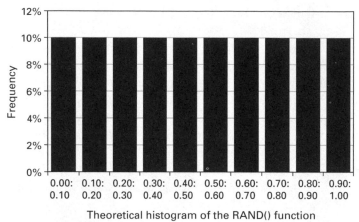

Theoretical histogram of the RAND() function

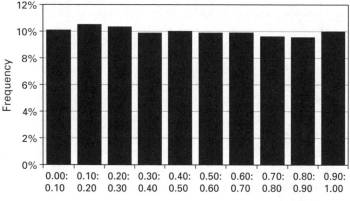

Empirical histogram of 10,000 trials of the RAND() function

Figure D.11
Theoretical and empirical histograms of the RAND() function.

keep in mind that working with some shape of the uncertainty is better than assuming that there is no uncertainty, as we do when working with base-case, deterministic valuation models. A deterministic input is equivalent to an extreme case of Monte Carlo simulation, where the histograms of all uncertainties consist of a single bar at the base-case value. Even a slightly spread-out distribution will be more realistic than the single bar.

Although a wrong distribution is better than no distribution, it is good practice to perform sensitivity analysis on the shapes. We can explore the effect of several distributional assumptions by running Monte Carlo using each of them. Figure D.15, for example, shows the difference

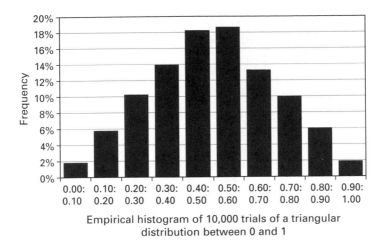

Empirical histogram of 10,000 trials of a triangular
distribution between 0 and 1

Figure D.12
Histogram of 10,000 realizations of a triangular distribution.

between using a uniform uncertainty (i.e., a simple RAND() function) and a triangular uncertainty for the uncertain inputs in the parking garage example. Different shapes of uncertainties do lead to differences in system performance. Extreme outcomes are less likely in the case of a triangular distribution, both along the vertical axis leading to a narrower cluster of clouds and along the horizontal axis, with more points clustered in the middle and fewer points at either end. However, the overall effect of the different distributions is, qualitatively, relatively mild. The uncertainty in system performance is significant whether one uses a uniform or a triangular distribution.

Generating Output Shapes

It is natural to think about the shape of the distribution of the uncertain performance. Calculating these output shapes from given input shapes is the essence of Monte Carlo simulation. Traditional spreadsheet models are "numbers-in, numbers-out" models, whereas Monte Carlo spreadsheets are "shapes-in, shapes-out" models.

We can generate the histogram of the outputs directly from the recorded results of the Monte Carlo simulation. Monte Carlo simulation add-ins can be useful in the construction of these graphical outputs. The facilities in Excel to create histograms are somewhat cumbersome.

A target curve, in contrast, is easy to build in Excel. This can be done by sorting the output of interest with values in ascending order and plotting the values against their respective percentiles, calculated from 0 to

Box D.4
Generating distributions in Microsoft Excel

Microsoft Excel has two random number generators: RAND() for a continuous variable between 0 and 1, and RANDBETWEEN for a discrete variable between any two specified integers. These functions allow us to generate a variety of other distributions. There are three main ways to do this.

Inverse Transform Method   Microsoft Excel has several inverse cumulative distribution functions (ICDFs). For example, the NORMINV function is the ICDF of the normal distribution, GAMMAINV is the inverse of the so-called Gamma distribution, and so on. We can use such functions to generate samples from the distribution using the RAND() function. In fact, if FINV is the ICDF of a random variable X with cumulative distribution function F, then "=FINV(RAND())" samples from the distribution of X.[10] In that sense, the RAND() function is the mother of all distributions. For example, the formula "=NORMINV(10,2,RAND())" generates a normal distribution with mean 10 and standard deviation 2. Likewise, the formula "=–A1*LN(RAND())" generates an exponential distribution with mean in cell A1.[11]

Sampling from a user-defined distribution   This approach uses a histogram, that is, a list of numbers (the mid-points of the histogram bins), and associated frequencies. The example in figure D.13 defines values in B2:B5 and associated probabilities in C2:C5.

We can combine the VLOOKUP and RAND() functions to sample from these values with the associated probabilities. To do this, set up a column with cumulative probabilities to the left of the value column B. This is done in A2:A5. It is important to start with 0 percent against the lowest value (i.e., the cumulative distribution values have a lag of 1, the 29 percent probability that the value is less than 20 is not put against the value 20 but against the next number up, the 50 in this case). Cell B8 then contains the formula "=VLOOKUP(RAND(),A3:B5,2)" and samples from the specified distribution.

Sampling from historical data   We can similarly sample directly from data of past occurrences of the uncertain input. For example, an important uncertain input for hospital operations is the length of stay. Suppose you have these data for the past 1,000 patients. Input it into a spreadsheet, labeling each record consecutively as in columns A and B in figure D.14. You can then sample from the data with the function "=VLOOKUP(RANDBETWEEN(1,1000),A2:B1001,2)," the formula in cell E3.

Sampling from historical data is only sensible if the process is reasonably stationary, that is, if the past is a good predictor for the future. For example, if you plot length of stay over time and observe that it tends to reduce, then you should modify the historical data to capture this trend before you can properly use it to sample future length of stay.

Box D.4
(continued)

	A	B	C
1	Vlookup values	Value	Probability
2	0%	10	12%
3	12%	20	17%
4	29%	50	37%
5	66%	100	34%
6		Total	100%
7			
8	Sampled value	120	

	A	B	C
1	Vlookup values	Value	Probability
2	0%	10	12%
3	=A2+C2	20	17%
4	=A3+C3	50	37%
5	=A4+C4	100	34%
6		Total	=SUM(C2:C5)
7			
8	Sampled value	=VLOOKUP(RAND(),A2:B5,2)	

Figure D.13
Sampling from a discrete distribution.

	A	B	C	D	E
1	Patient	Length of stay	Length of stay sample		
2	1	1	2		
3	2	2			
4	3	3			
5	4	4			
6	5	5			
7	6	6			
8	7	7			
9	8	8			
10	9	9			

	A	B	C	D	E
1	Patient	Length of stay	Length of stay sample		
2	1	1	=VLOOKUP(RANDBETWEEN(1,1000),A2:B1001,2)		
3	2	2			
4	3	3			
5	4	4			
6	5	5			
7	6	6			
8	7	7			
9	8	8			
10	9	9			

Figure D.14
Sampling from historical data.

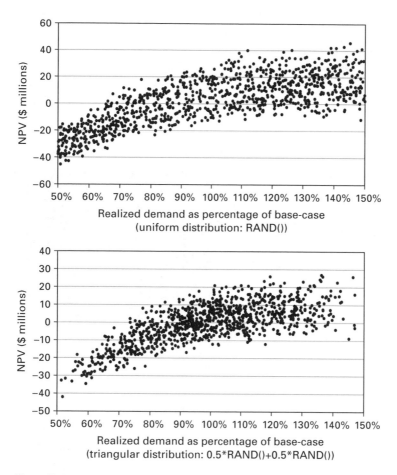

Figure D.15
NPV distribution as a function of deviation from demand projection for uniform uncertainties (top) and triangular uncertainties (bottom).

1 in ascending steps of 1/n if there are n sorted output values. Alternatively, we can use the PERCENTRANK function to calculate the percentage rank for each output within the complete set and then scatter-plot the output against its percentage rank, as figure D.16 illustrates.

Step 5: Introduce Dependence between Uncertain Numbers

This step recognizes the relationships between uncertain inputs, which can significantly affect the shape of the distribution of the output,

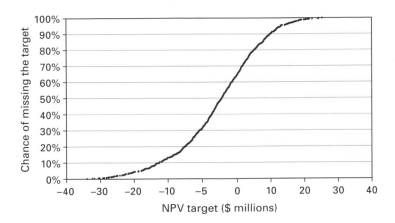

Figure D.16
Target curve generation via scatter plot of output against its percentage rank.

as box D.5 indicates. It is the most difficult step. Although it is relatively easy to acknowledge the existence of relationships, quantifying them in a model is hard, not least because our intuition about the nature of the relationships between uncertainties is often relatively poor.

In the parking garage example, one can assume that demand and annual revenue per space are related. If demand is high, the operator of the parking garage may be able to charge more per space, which will increase the annual revenue per space. This results in a positive relationship between revenue per space and demand. There are two ways to model this relationship: either as a joint distribution directly or by modeling the driving mechanism for the relationship. For example, in the case of the parking garage, we could capture the driving mechanism with a pricing model that inputs demand over time, calculates appropriate prices we would charge, and thereby produces revenues per space as a function of demand.

A direct model of the relationship could be of the form:

Average revenue/ space = sampled annual revenue/ space
    + b*demand deviation from projection,

where b is a parameter that needs to be determined sensibly. If we choose b = 0, then the average revenue per used space is sampled independently of demand, as in step 4. The second factor corrects this by taking demand

Box D.5
Why relationships between uncertain inputs matter

Relationships between input variables can significantly affect the shape of the output of models. To illustrate this, consider the simplest of all models, the sum of uncertain inputs. Note that the expectation of the sum of uncertain summands is the sum of their expectations regardless of the relationships between them. However, although the average of the histogram of the model output remains fixed when we introduce relationships, its shape can change dramatically.

When we add up unrelated uncertain numbers, the shape of the sum will be more peaked in the middle than the shape of the summands.[12] To get an extremely high or low sum, we have to be lucky enough to sample only high or only low summands. Sums in the middle are more likely because they can be a sum of middle-sized summands as well as mixes of ups and downs of individual summands. The histogram of a sum peaks in the middle.

This peaking effect is affected when the summands are related to each other. To see this, consider the sum of two spreadsheet cells, A1 containing the formula "=RAND()," and A2. Cell B1 contains the formula "A1+A2." The content of A2 is a random number more or less related to A1.

• Suppose first that A2 also contains the formula "=RAND()" and is independent of A1. When you do a Monte Carlo simulation, you will find that the shapes of the uncertain numbers in A1 and A2 are both flat but that the shape in B1 is triangular, that is, it peaks in the middle as expected.

• Now suppose that A2 "=A1." Both A1 and A2 contain uncertain numbers and their shapes are both flat. However, the shape of their sum in B1 is now also flat, a stark difference from the previous triangular shape. The reason is that the uncertain numbers in A1 and A2 are now perfectly positively related. The mixing of high and low realizations of A1 and A2 to get a result in the middle no longer happens. If A1 is high, so is A2. In this extreme case, the peaking completely disappears. In less extreme cases, when there is a less than perfect positive relationship, the peaking abates.

• Finally, suppose A2 "=1 – A1." A1 and A2 are flat shapes as before. However, the shape of B1 is now extremely peaked. B1 now always is 1 no matter what. The uncertain numbers in A1 and A2 have a perfect negative relationship. Whenever we sample a low A1, it is balanced out by a correspondingly high A2 that leads to the sum of 1. In the extreme case of a perfectly negative relationship, the peaking becomes maximal. In less extreme negative relationships, the sum exhibits a stronger peaking than in the case of independent summands.

In summary, the shape of the sum of uncertain numbers peaks more than the shape of the summands. Negative relationships between the summands amplify this peaking effect, whereas positive relationships weaken it.

Box D.5
(continued)

Relationships between input variables can also affect the average output, thus inducing a Flaw of Averages. Take, for example, the expression: net revenue = margin * sales volume. If margin and sales volume are unrelated, then the average revenue equals the product of average margin and average sales volume. However, if margin and sales are related, as one would expect because demand drives them both, then this identity fails. If demand for a patent-protected product is higher than expected, then the company can charge higher prices and obtain both higher margins and higher sales volume—a positive relationship. However, if demand for an unprotected product is higher than expected, then this can lead to more competitors in the market, fiercer price competition, and depressed margins. Meanwhile, the consolidated marketing effort of all competitors may increase market size and higher than expected sales volume. This dynamic induces a negative relationship between sales volume and margin.

When margin and sales volume are positively related, the expected net revenue will be larger than the product of expected margin and expected sales volume. If the relationship is negative, then the inequality is reversed and expected net revenue is smaller than projected. To illustrate this effect in a spreadsheet, take the distribution of margin in cell A1 as "=RAND()" and set sales volume in cell A2 as "=A1+RAND()" for a positive relationship, in cell A3 as "=2-A1-RAND()" for a negative relationship, and in cell A4 as "=rand()+rand()" for no relationship. In all three cases, the individual distributions of margin and sales volume are the same; margin is uniformly distributed between 0 and 1, and sales volume has a triangular distribution between 0 and 2. However, both the shape and the average of the distribution of net revenue, the product of the respective cells, are quite different.

---

deviation from projection into account. So what should b be? First, b should be positive because intuitively the relationship between the two inputs is positive (i.e., in high-demand scenarios, the price will increase and raise the revenue per used space). Second, parameter b reflects the price elasticity of demand, which is itself an uncertain input in the model. It should therefore have a distribution rather than a single value. If there are data on price elasticity of demand for parking garages, these may be useful in estimating this distribution.

Notice that the above model does not change the average revenue per space as long as the price elasticity parameter b itself is independent of the demand deviation from projection. This is because the average demand deviation from projection is zero.[8]

Common Undercurrents Cause Relationships between Uncertain Inputs

Relationships between uncertain inputs are often a consequence of common undercurrents—of global drivers further up the causal chain that are not directly included in the model. For instance, the state of the economy, characterized by metrics such as GDP growth, is a common undercurrent that affects the performance of most systems. If the economy thrives, then both the demand for our parking garage and its operating costs may go up. This induces a positive relationship between costs and demand.

It is useful to explore such common causes. First, make a list of the key undercurrents that might simultaneously affect the uncertain inputs. Then capture the qualitative nature of this effect in a matrix, as figure D.17 illustrates. The columns correspond to undercurrents, and the signs in the matrix denote the anticipated effect on the uncertain input as "higher than expected" (+) or "lower than expected" (−), respectively.

An undercurrent matrix is a useful tool to start a discussion about relationships between uncertainties. Its development is a pragmatic rather than a scientific exercise. Although data can and should be used as much as possible, the critical challenge is prioritization—the determination of a manageable list of key undercurrents from a vast number of possible variables that may affect the uncertain inputs. This requires expert judgment and context knowledge.

Once a list of key undercurrents is determined, we can use them as additional, hidden inputs in our model and determine the uncertain inputs y_i via equations of the form

$$y_i = a_{i0} + a_{i1}x_1 + \ldots + a_{in}x_n + e_i,$$

	Economy grows faster than expected	Inflation increases	Tighter immigration rules are introduced
Demand increase	+	−	−
Unit revenues increase	+	+	
Operating costs increase	+	+	+

Figure D.17
Undercurrent matrix for uncertain inputs for the parking garage model.

where $x_1, \ldots x_n$ are the undercurrents and e_i is a residual uncertainty that accounts for uncertainty in the variable y_i on top of the undercurrents. Such models can easily be implemented in a Monte Carlo spreadsheet provided we have estimated coefficients $a_{io}, a_{i1}, \ldots, a_{in}$, and agreed-on distributions for the inputs $x_1, \ldots x_n$ and $e_1, \ldots, e_m$. The variables $y_1, \ldots, y_m$ are then calculated using the above formula. It is appropriate to seek statistical advice at this point.

A word of caution is in order. It would be wrong to get the impression that relationships between undercurrents and uncertain inputs can be straightforwardly established. This can be a lengthy debate, hopefully informed by research and data. The relationship may well not be easily captured by a + or a – sign, let alone a linear equation as assumed above. However, neglecting relationships can be worse than getting their magnitudes slightly wrong.

This step in the process requires more statistical and econometric expertise than the other steps. It is advisable to experiment first with a range of intuitively sensible relationship models without worrying too much about whether they are realistic. If these do not change the results much, you have a robust situation and may just pick one of them without much concern. However, if the relationship matters and changes the results drastically, then you should spend the time and effort to choose an appropriate model. This is difficult. If you lack statistical experience, you should seek professional guidance at this point.

Step 6: Introduce Dynamically Changing Uncertain Numbers

For the modeling of flexibility, it is important to acknowledge explicitly that many uncertain inputs to the system model will evolve over time and that we can use the flexibility as the uncertainties unfold. We therefore need to spend some time on dynamic models of uncertainty.

As appendix E discusses, the classical dynamic model of uncertainty is a random walk. The simplest example of a random walk is a repeated coin flip, for example, where you pay $1 when the coin shows head and gain $1 when it shows tail. The random walk keeps track of your profit and loss over time. In mathematical terms, a simple random walk is a sequence of uncertain numbers where $X(t)$ evolves from $X(t-1)$ by adding an uncertain shock $\varepsilon(t)$

$$X(t) = X(t-1) + \varepsilon(t),$$

where $X(0)=x_0$ is known and $\varepsilon(t)$ are independent uncertain numbers, typically with the same distribution. In the example of the coin flip, $x_0=0$ and $\varepsilon(t) = 1$ with probability 1/2 and $\varepsilon(t) = -1$ with probability 1/2.

Random walk models will often provide a more realistic view of future scenarios than fixed evolution curves defined by a fixed set of parameters that we sample randomly to account for uncertainty. Consider uncertainty in demand growth for the parking garage example. In step 3, we modeled its uncertainty by sampling a random deviation within ±50 percent of the base-case demand and then calculating the corresponding growth curve. The scenarios were thus all within a limited band around the base case, which implied a rather smooth growth projection. It is likely that the actual growth will be much more uneven. To simulate the more realistic situation, we can use a random walk model. The initially sampled growth curve will only give us the trend, and the random walk will modulate around this trend.

Random walks are easy to implement in spreadsheets. We begin by sampling the smooth growth curves, giving expected demands d_1, d_2,..., d_{15} over the 15-year planning horizon. Then we start the random walk process with $X(1) = d_1$ and use the recursion

$$X(t) = X(t - 1) + \varepsilon(t), t = 2, 3,...,15,$$

where $\varepsilon(t)$ is a random shock with an expected value of $d_t - d_{t-1}$.[9] This guarantees that the expected value of $X(t)$ is d_t. However, $X(t)$ fluctuates around this expected value. Figure D.18 shows ten realizations of demand paths, with normal random shocks, and illustrates how the realizations

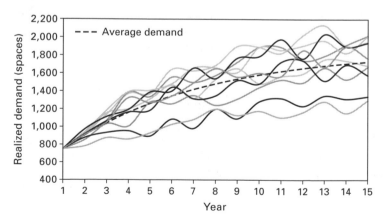

Figure D.18
Realizations of a random walk model for demand.

modulate around the average growth curve. It also shows that the demand realizations spread out over time, that is, uncertainty in demand grows with increased time. This is sensible. It is more difficult to predict the more distance future. Notice that this was not the case in our original model, which only modeled the average growth curves. The models can of course be combined by first generating an average growth curve and then a random walk modulation around this curve. Appendix E provides more details on the specifications of dynamic input distributions.

Step 7: Using Rules for Exercising Flexibility

Steps 1–6 turned a fixed-number model with base-case inputs into a Monte Carlo simulation model that allows us to calculate with shapes. It converts input shapes into output shapes and allows us to explore the effects of uncertainty, of the variation around the base values as well as of the dependence between variables. Such probabilistic models are necessary for a systematic valuation of flexibility because we only exercise flexibility in certain scenarios. If we do not have a way to simulate the scenarios, the value of flexibility is invisible.

To value flexibility within a probabilistic model, we have to tell the model when to use this flexibility. This is the last part of the modeling exercise. In the parking garage example, we need to tell the model when to expand and by how much. A simple way of doing this is to use the Excel IF function. For example, we could stipulate that we should add an extra level to the garage if it ran at its effective capacity for the past 2 years. To do this, we simply need to keep track of capacity utilization, year on year, and the IF function would trigger the addition when demand met the stated conditions.

Figure D.19 shows a modification of the NPV spreadsheet in figure D.1, which includes such a rule for exercising flexibility. The row labeled "Expansion?" contains the IF function statements. We have not included expansion in year 1 or year 15 because we could not meet the condition in the first year and would not want to expand in the last year. For example, the cell in the "Expansion?" row corresponding to year 3 (cell E6) contains the formula

"=IF(AND(D3<MAX_CAP,MIN(D4,D5)
+MIN(E4,E5)=D5+E5),"expand","")."

This statement has two conditions. The first is that the number of levels in year (cell D3) is not yet at its maximum; MAX_CAP is a name

Performance Calculation							
Year	0	1	2	3	4	...	15
Garage levels		4	4	4	5	...	8
Realized demand (spaces)		750	893	1015	1120	...	1634
Capacity (spaces)		800	800	800	1000	...	1600
Expansion?				expand	expand	...	
Build extra capacity (spaces)		0	0	200	200	...	0
Revenue ($ M)		7.5	8.0	8.0	10.0	...	16.0
Operating costs ($ M)		2.4	2.4	2.4	3.0	...	4.8
Land leasing costs ($ M)	3.3	3.3	3.3	3.3	3.3	...	3.3
Expansion cost ($ M)	0.0	0.0	0.0	4.5	5.0	...	0.0
Cashflow ($ M)	-3.3	1.8	2.3	-2.3	-1.3	...	7.9
DCF ($ M)	-3.3	1.6	1.9	-1.7	-0.9	...	1.9
Present value of cashflow ($ M)	20.4						
Capacity cost for up to two levels ($ M)	8.8						
Capacity costs for levels above 2 ($ M)	7.9						
Net present value ($ M)	3.7						

Figure D.19
Spreadsheet with decision rule for expansion.

assigned to a cell that contains the maximum number of floors. The second condition guarantees that demand during the past 2 years was above the garage's effective capacity. When the conditions occur, building cost is incurred (row labeled "Expansion cost"), and the additional capacity becomes available the following year (row labeled "Capacity" in year 4, cell F5). In this case, the rule for exercising flexibility stipulates that we build only one level at a time. Because demand grows rapidly in the scenario sampled in figure D.19, it leads to expansions in both years 3 and 4. At the end of the planning horizon in year 15, the garage has been expanded to eight levels.

With a rule for exercising flexibility in place, the performance of the garage depends not only on the design parameters, such as the initial number of levels, but also on our choice of the rule. We complement the optimization of today's actions with an optimization of our anticipated future actions. The technical term for this is dynamic programming. Notice that it is important that the rule is only based on information available at the time of the decision. A rule that would exercise expansion in year 4 based on demands in years 5 and 6 is not allowed.

Final Comments

Expectation Consistency

When we turn a static base-case model into a probabilistic model for Monte Carlo analysis, we have to make sure that we remain expectation consistent, i.e., that the expected values for our inputs stay fixed at the base-case values. Otherwise, we start comparing apples with pears. If, for example, we implemented a demand model that would lead to higher demands, on average, over the time horizon of the parking garage, then it would not be surprising to see that the parking garage is worth more, on average, based on the probabilistic model instead of the static model.

To check expectation consistency, analysts should record both the model outputs of a Monte Carlo simulation and all randomized inputs. They should then compare the input averages with the corresponding base-case values. Expectation consistency requires these values to be close to one another.

Trials Needed for a Robust Monte Carlo Simulation

The goal of a Monte Carlo simulation is to perform a shape-in, shape-out calculation—to approximate the distributions of uncertain outputs given distributions for uncertain inputs. It is important to understand that Monte Carlo simulations only provide approximate results. They could only provide precision after an infinite number of trials. The more trials, the more accurate the approximate result—but it will never be precise. Therefore, a critical question is: How many trials do we need for practical purposes?

To get a first idea of the accuracy of the Monte Carlo process, we can calculate the accuracy of our estimate of the mean of the distribution of the output. Clearly, a good approximation of the mean is a necessary condition for a good approximation of the whole distribution. This approximation of the mean is the average of the generated output trials. An important statistic for its accuracy is its standard error, which is the standard deviation of the generated output trials divided by the square root of the number of trials. Elementary statistics tells us that the following formula defines the 95 percent confidence interval (CI) for the mean

95 percent CI = Average + 2 * Standard Error.

What does a 95 percent CI signify? Suppose we repeat Monte Carlo simulation with the same number of trials again. Because we sample inputs randomly, we will obtain a new estimate of the average and also a new standard error and a new 95 percent CI. We chose the boundaries of these 95 percent CIs so that we can expect 19 out of 20 such repeated Monte Carlo simulations to produce CIs that contain the actual (unknown) mean of the distribution of the uncertain output. Thus, a minimum requirement for accuracy is that the 95 percent CI is small, that is, that the standard error of the mean is small relative to the mean. If this is not the case, we need to run more trials.

Note that an accurate estimate of the mean is not a sufficient criterion for an accurate estimate of the distribution. To estimate the probability that the output value falls into any specific region within its range, for example, into a specific bin of its histogram or below a certain target value, we can again use the standard error of the mean. This probability is the average of a "counting" random variable obtained directly from the output, assigned the value 1 if the sampled output value falls in the specified region and the value 0 if not. We estimate a 95 percent CI for this probability using the standard error of the counting variable. If P is

	A	B	C	D
1	−2.99960	0.01	0.000	0.030
2	−2.74998	0.02	0.000	0.048
3	−2.17415	0.03	0.000	0.064
4	−2.06954	0.04	0.001	0.079
5	−1.94391	0.05	0.006	0.094
6	−1.80161	0.06	0.013	0.107
7	−1.68137	0.07	0.019	0.121
8	−1.56787	0.08	0.026	0.134
9	−1.53898	0.09	0.033	0.147
10	−1.43279	0.10	0.040	0.160
11	−1.41880	0.11	0.047	0.173
12	−1.20386	0.12	0.055	0.185
13	−1.16213	0.13	0.063	0.197

Figure D.20
Calculation of confidence bounds on target curve. Column A: first 13 of 100 observations sampled from a standard normal variable, sorted in ascending order. Column B: target curve value P associated with sorted observations, ascending in steps of 1/100. Column C: calculation of lower confidence bound, for example, C1: "= max(0,B1-2*SQRT(B1*(1-B1)/100))." Column D: calculation of upper confidence bound, D1="min(1,B1+2*SQRT(B1*(1-B1)/100))."

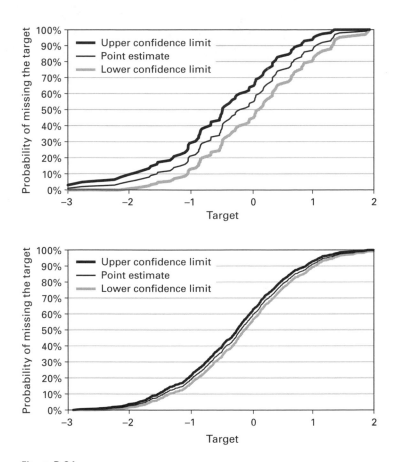

Figure D.21
The 95 percent confidence bounds on target curve sampled from a standard normal distribution (mean 0, standard deviation 1) with 100 trials (top) and 1,000 trials (bottom)

the proportion of trials that fall into the region, then the standard error of the counting variable is given by the square-root of $P*(1 - P)/n$, where n is the number of trials. Figures D.20 and D.21 illustrate the calculation of confidence bounds for a target curve.

Appendix E: Dynamic Forecasting

Current forecasting practice produces single-number (or "point") projections for the future. Such projections, suggesting that it is possible to pinpoint the future, are unrealistic. They do not indicate the level of uncertainty appropriate for the forecast. Thus, they are certainly inadequate for a systematic appraisal of flexibility.

Instead of point predictions, we need a practical way to present the uncertainties around forecasts. Our suggestion is that we should use dynamic forecasts. These are spreadsheet modules of the uncertain environment in which the system operates. This context is characterized by the evolution of uncertain variables, such as demand, costs, prices, and productivity. Dynamic forecasting spreadsheet modules are scenario generators for these variables. They implement a joint distribution of these variables, including dependencies and their evolution over time, where appropriate. We link these modules to valuation models to drive the uncertain input variables.

An example of a dynamic forecast is the demand module that drives the parking garage case of chapter 3 and appendix D. The purpose of this appendix is to introduce the main models that are used for such dynamic forecasts and to illustrate how to:

- Build forecasting modules in spreadsheets, and
- Calibrate the modules to historical data.

Random Walk Models

The simplest model of an uncertain variable that evolves over time is a series of coin flips, resulting in a sequence of heads and tails, such as HTHHTTHTTHT. The model is particularly simple because the variable can only take on two values but also, and more importantly, because the

variable in period t does not depend on the variables in earlier periods
t − 1, t − 2, and so on. This is an unusual situation. For most parameters
of interest—such as demand, price, productivity, and others—their level
at any time will depend on the past.

A stock-flow model is the simplest model of a variable that depends
on the past. It is a direct extension of the coin flip model. The variable
of interest is the stock, that is, the amount of some variable aggregated
over time. In the financial flow version, we focus on the amount of money
gained. In that case, the coin flip leads to either a specified gain or a loss,
for example, with a head leading to a gain of $1 and a tail leading to a
loss of $1. This model is of the form

$$X(t) = X(t - 1) + \varepsilon(t),$$

where $X(0) = 0$ and $\varepsilon(t)$ is the financial flow driven by consecutive independent coin flips, for example, $\varepsilon(t) = -1$ with probability ½ and $\varepsilon(t) = 1$
with probability ½. This model is simple to implement in a spreadsheet
as figure E.1 illustrates.

The stock flow model is versatile: Its random component $\varepsilon(t)$ can have
any distribution. For example, if we choose $\varepsilon(t)$ to be a normal random
variable, then we obtain the so-called random walk model (also known
as arithmetic Brownian Motion). We may also wish to model the fact
that some factor of $X(t)$ carries over to the next period, which leads to
a model of the form

	A	B	C
1	Time	Flow	Stock
2	0		0
3	1	1	1
4	2	1	2
5	3	1	3
6	4	−1	2
7	5	1	3
8	6	1	4
9	7	−1	3
10	8	−1	2

Figure E.1
Stock flow model implemented in a spreadsheet. Cell B3 contains the formula =if(RAND()
< 0.5,–1,1), which generates 1 with probability ½ and –1 with residual probability ½. Cell
C3 contains the formula =C2+B3, that is, the generated flow is added to the stock of the
past period. The formulas in B3 and C3 are then copied down in B4:C10. Recalculating the
spreadsheet by pressing the F9 key produces successive new scenarios of stock level over
time. Note that the stock can become negative in this model, recording losses.

$X(t) = a*X(t-1) + \varepsilon(t),$

where the factor "a" captures a nonrandom rate of appreciation (a>1) or depreciation (a<1).[1] For example, future demand might be growing at the rate of r = 5 percent (thus, a = 1.05) with some variation $\varepsilon(t)$ around this trend.

Multiplicative models are an alternative to additive models. In this case, the error $\varepsilon(t)$ is added to a scaled past period level, and the model is of the form

$X(t) = \varepsilon(t) X(t-1)^a.$

When the exponent a = 1, this model is called a multiplicative random walk. It can be thought of as a generic model of a bank account $X(t) = X(t-1) + (\varepsilon(t) - 1)X(t-1)$ with a randomly fluctuating interest rate $r(t) = (\varepsilon(t) - 1)$. A multiplicative model reflects random growth, proportional to the existing amount in stock. If we start with a positive initial level $X(0)$ and apply positive growth rates $\varepsilon(t)$, then future amounts $X(t)$ will never be negative. This is a useful feature for many kinds of variables, such as the demand for or price of some asset, which are never negative.

Note that the multiplicative model turns into an additive model after a log-transformation:

$\log(X(t)) = a*\log(X(t-1)) + \log(\varepsilon(t)).$

A typical assumption for the distribution of the error term $\varepsilon(t)$ of a multiplicative model is that it follows a log-normal distribution, that is, $\log(\varepsilon(t))$ is normally distributed.

The simplest models are of the form $X(t) = X(t-1) + \varepsilon(t)$ or $X(t) = \varepsilon(t)X(t-1)$, where $\varepsilon(t)$ has only two values, that is, $\varepsilon(t) = u$ with probability p and $\varepsilon(t) = d < u$ with complementary probability 1-p. These models are (additive or multiplicative) lattice models because the values change in a discrete fashion as if moving from a vertex to one of the neighboring vertices in a lattice. If we start from some value $X(0) = x_0$ in the additive model, $X(1)$ can only move to either $x_0 + u$ or $x_0 + d$. $X(2)$ can therefore only achieve one of three values, $x_0 + 2u$ (twice up), $x_0 + u + d$ (once up, once down), or $x_0 + 2d$ (twice down). These models are thus said to be "recombinant" in that possible outcomes combine in each stage, as in the case of "once up, once down" = "once down, once up." They thus have the great advantage that the number of possible end points after N stages is (N + 1) rather than 2^N, which is the number of possible states after N stages if the process moves into one of two nonrecombinant states in each stage (e.g., if $\varepsilon(t)$ has two possible realizations $u(t)$ and $d(t)$, which change with t).[2] In the case of the recombinant lattice, $X(t)$ can have t+1 values $x_0 + su + (t-s)d$, where s ranges from 0 to t and assumes a binomial distribution over these values, with success probability p. The same principle applies to the multiplicative model, where $X(1)$ would have values x_0u or x_0d, $X(2)$ can have values x_0u^2, x_0du, or x_0d^2, and so forth.

Calibration of Dynamic Forecasting Models

We can easily implement dynamic forecasting models in spreadsheets, analogously to figure E.1. Before doing so, we need to make sure we are doing the right thing. We need to address two questions:

- Are the models appropriate?
- What values should we choose for the unknown parameters?

Historical data and statistics help with these questions. A detailed exposition goes beyond the scope of this appendix, but an example illustrates the main principles.

Before we explain the example, let us be clear about what statistics can and cannot do for us. Statistics alone cannot determine an appropriate model; we can only use statistics to test the degree to which a model is consistent with observed data. If the model is consistent with the data to a high degree, that does not mean that the model is appropriate; it only means that the data do not allow us to refute the model. In general, many models, often with contradictory implications, can fit a given set of data. This is particularly the case when we build models to capture the relationship between variables based on sets of data over time. This is because many phenomena exhibit monotone growth, either upwards or downwards. Such steadily growing and declining variables correlate naturally, even if there is no causal relationship at all between them. Statistics by themselves are not sufficient to determine an appropriate structural model that captures the relevant causal relationships. Its development is a task for a domain expert who understands the relationships between different factors.

It is quite often the case that several models are consistent with domain expertise and the available data. This issue is model uncertainty and is particularly relevant when different models produce significantly different ranges of future scenarios. In such situations, we should work with several models simultaneously. We implement our spreadsheet to generate a new future scenario by first choosing a model at random from a list of appropriate models and then use this model to generate the scenario. We then use a range of models to generate the resulting range of scenarios. When model uncertainty is relevant but not taken into account, the uncertainty produced by the spreadsheet module will be lower than the real uncertainty faced by the system.[3]

To illustrate how we would go about determining an appropriate model, consider the data in figure E.2, which represent an uncertain variable, say demand for some service or product, over a period of 35 months. These data do not come from a real-world process; we have simulated them from a predetermined process. The reason we did this is to make the point that we often put too much meaning into data, when the actual cause of the data is random variation.

It is quite easy in many practical circumstances to develop a convincing story that "explains" a set of data. Suppose in this case that the data are orders of a new product that a company introduced 35 months ago. The "explanation" of the pattern of sales might go as follows:

• Initial success due to significant prelaunch marketing;

• Sales fluctuation over the next 10 months ("Our competitor started reacting to the new product with mixed success");

• A more stable period with somewhat lower sales ("Our competitor and the market adjusted to the new product");

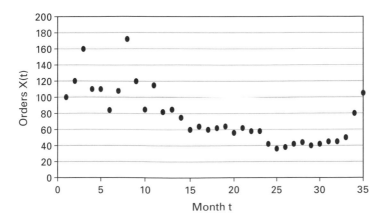

Figure E.2
Monthly data series for 35 consecutive months.

• A decline to a low in month 25 ("Our competitor launched their own new product"); and

• Significant recovery ("We hired a new chief technology officer who turned things around").

As convincing as such a story might be, purely chance fluctuations can often explain historical data equally well (as in this case).

Statisticians are the advocates of luck. They start from the premise that an apparent pattern like the one in figure E.2 is just a fluke. They then use "structural models," specific functional forms, like the random walk model, that include randomness and "fit" these models to the data to see how well they explain the data and how much of the variation remains "unexplained." Two things are important in relation to the choice of structural models: that the models make sense to domain experts and are simple.

Simplicity of structural models is important for two reasons. One is that simple models are easier to explain to a wide audience. Another, less appreciated, rationale for simple models is that they typically depend on fewer parameters. Complex black box models with many parameters to estimate are much less appealing. They tend to fit well but produce worse predictions than simpler, less "accurate" models. This phenomenon is called "overfitting." The situation becomes worse when your model uses "data" that are not directly observed but result from another forecasting exercise, for example, if you predict future demand by first predicting GDP growth. The forecast error in GDP prediction is likely to exacerbate the forecast error of demand.[4]

Returning to the example data in figure E.2, consider the possibility that chance generated the data according to the additive forecasting model described above: $X(t) = a*X(t-1) + \varepsilon(t)$. We can fit this model by scatter-plotting $X(t-1)$ against $X(t)$ and performing a regression to estimate a value for the unknown parameter a. Figure E.3 shows the result. The scatter plot is not convincing, however. It shows somewhat irregular behavior. Many points cluster close to the lower end of the line. The variation parallel to the y-axis does not appear to be the same at the upper and lower ends (the upper end is larger). This phenomenon is heteroscedasticity (a fancy word meaning "different variation"). It renders linear regression problematic because it implies that the error term is not from the same normal distribution, which is a general assumption of regression analysis. In practical terms, it means that the large variation points have an undue weight on the position of the line.

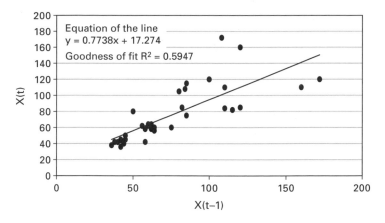

Figure E.3
Regression model for $X(t) = aX(t - 1) + \varepsilon(t)$.

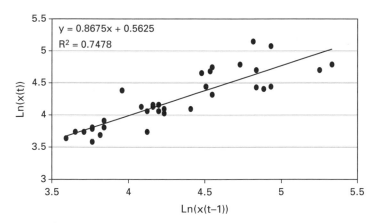

Figure E.4
Regression model for $Ln(X(t)) = a*Ln(X(t - 1)) + \varepsilon(t)$.

One way to correct for heteroscedasticity is to work with a different model, one that dampens the greater effect of higher values. The multiplicative model discussed above is such a model. To obtain it, we perform a log-transformation of the variables. We can then estimate the unknown parameter by regression analysis as before. Figure E.4 shows the result. This model is a better fit to the data. We specify it as

$$Ln(X(t)) = 0.87*Ln(X(t - 1)) + 0.56 + \varepsilon(t),$$

or, equivalently,

$X(t) = X(t - 1)^{0.87}*\exp(0.56 + \epsilon(t))$.

What would be a sensible distribution for the random perturbation $\epsilon(t)$? This distribution is best approximated by the historical errors. The distribution of the errors has a mean of zero, and we calculate its standard deviation as 0.22.[5] Because we have a relatively small set of errors, it is difficult to ascertain that the errors are normally distributed just by looking at the histogram. The comparison of the empirical target curve of the data and the associated normal distribution with mean 0 and standard deviation 0.22 casts some doubt on the appropriateness of the normal distribution assumption for this model (see figure E.5), and specific statistical testing of this assumption seems in order.[6] An additional assumption that needs to be checked is whether we can assume the errors to be statistically independent. Figure E.6 provides no evidence to suggest that the errors are statistically dependent.

As a first working model, the specification $X(t) = X(t - 1)^{0.87}*\exp(0.56 + \epsilon(t))$ with independent normal errors $\epsilon(t)$ with mean zero and standard deviation 0.22 seems a reasonable model that might have generated the data in figure E.2, provided that the multiplicative forecasting model was sensible in the first place. The actual model that we used to generate the data was $X(t) = X(t - 1)^{0.9}*\exp(0.1 + \epsilon(t))$, where $\epsilon(t)$ are independent shock terms with a normal distribution with mean zero and standard deviation 0.2.

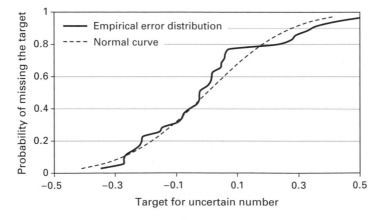

Figure E.5
Empirical cumulative distribution function of the regression errors versus normal distribution.

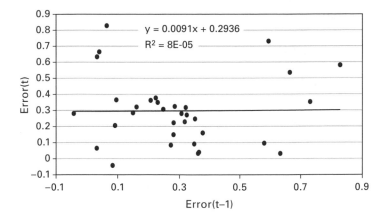

Figure E.6
Scatter plot of successive errors. The line slope is not statistically significant, and the correlation, as measured by R2, is practically zero.

Note that we have calibrated the model on the same data that we use to test it. This is common practice, but it is statistically flawed. In the ideal situation, you would use one set of data to estimate the parameters and another set to test model consistency with the data. In practice, analysts rarely do this because they lack sufficient data. However, if you have sufficient data, we recommend that you split the data into one set for model calibration and another for model testing.

Seasonality

Random walks of the form $X(t) = X(t - 1) + \varepsilon(t)$ with independent and identically distributed error terms $\varepsilon(t)$ have a fixed trend in the form of the expected error. This fixed trend is sometimes made explicit by writing $X(t) = X(t - 1) + b + \varepsilon(t)$, where the associated error has a mean of zero (as in the above calibration for the example data). Such models can be extended to models of the form $X(t) = X(t - 1) + b(t) + \varepsilon(t)$, where $b(t)$ captures some known seasonal effects. For example, if seasonal effects are assumed to be quarterly, the function $b(t)$ can be chosen of the form $b(t) = b_0 + b_1 * Q_1 + b_2 * Q_2 + b_3 * Q_3$, where $Q_i = 1$ if t lies in quarter i of the year and $Q_i = 0$ otherwise. We can determine the coefficients using multivariate regression, with so-called dummy variables for the quarters.

Mean Reversion Processes

Random walks have the unpleasant property that they can "blow up" over long time horizons, meaning that they project unreasonably large or small values for the long term. When the model has moved to a level $X(t)$ at time t, then, given this position, it does not know about or has "forgotten" about the old mean $X(0)$. The new mean of $X(t + s)$ given $X(t)$ at time t is now $X(t)$, not $X(0)$. This forgetting about the mean, about the "natural home" of the process, can lead to seriously high or low values over time, to the extent that the generated paths look unrealistic to experts. Put another way, if we are at $t = 0$ and a process evolves according to the model $X(t) = X(t - 1) + \varepsilon(t)$ where $\varepsilon(t)$ has mean zero and variance σ^2, then $X(t)$ is a random variable with mean $X(0)$ and variance $t*\sigma^2$. From today's perspective, the mean does not change for longer time horizons, but the variance increases linearly. Figure E.7 shows this funnel effect.

A way to avoid this often unrealistic situation is to use a model that tends to center on a long-term mean value, that is, a mean reversion model. To create this effect, we augment a model with a mean reversion term: $r*(m - X(t - 1))$, where m is a fixed number, the long-term mean or "natural home" of the process, and r is the fixed rate of mean reversion between 0 and 1. Mechanically, we can think of this as a damping effect: If $r = 0$, there is no damping or mean reversion, whereas if $r = 1$, then the process resets to the mean each period.

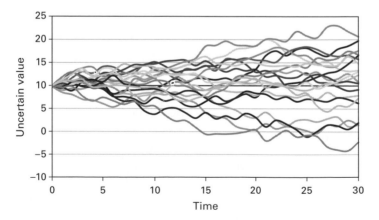

Figure E.7
Twenty paths of an additive random walk $X(t) = X(t - 1) + \varepsilon(t)$.

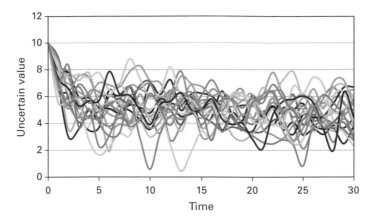

Figure E.8
Twenty paths of a mean reversion process $X(t) = X(t-1) + 0.5*(5 - X(t-1)) + \varepsilon(t)$.

For example, the mean reversion modification of the additive random walk model has the form:

$X(t) = X(t-1) + r*(m - X(t-1)) + \varepsilon(t)$.

The mean reversion term $r*(m - X(t-1))$ moves $X(t-1)$ somewhat towards m before the random shock $\varepsilon(t)$ is added. This correction towards the mean before a random error term is added implies that the process becomes less likely to move too far from m. Moreover, the further away $X(t-1)$ is from m, the larger the correction $r*(m - X(t-1))$. Figure E.8 illustrates this tendency to move towards the mean in contrast to the random walk in figure E.7.

To calibrate the above mean reversion process, we can scatter $X(t)$ and $X(t-1)$ and regress $X(t)$ on $X(t-1)$ as before. This will give a model $X(t) = a*X(t-1) + b + \varepsilon(t)$. We then match the coefficients with $X(t) = r*m + (1-r)*X(t-1)+\varepsilon(t)$, that is, $r = 1 - a$, and $m = b/(1-a)$.

Models with Jumps

Random walks and mean reversion models change incrementally. Large deviations from $t-1$ to t, although possible, are unlikely. It sometimes makes sense to include the possibility of large deviations possibly due to special events such as wars, regulation changes, or other disruptions. We typically do this by overlaying a "jump process" onto a smoother process, such as a random walk.

A convenient way to model a jump process is by modeling the random time between two large deviations. Random durations are often modeled using an exponential distribution. Figure E.9 shows that the exponential distribution has a sensible shape for random durations.[7] The distribution can be easily implemented in Excel (see box D.4). All that is required is an estimate of the mean duration, the expected time between disruptive events.

The exponential distribution has an interesting and unique property, called lack of memory. It means that the chance of something happening does not depend on the length of time you may have already been waiting. In technical terms, the probability of having to wait until time (t + s), given that you have already waited until time t, is the same as the probability of having to wait until time s from the start.[8] We can therefore at every time period t = 1, 2, 3... draw a sample waiting time to the next event from an exponential distribution. If the waiting time generated at time t – 1 leads to an event before time t, then we add an additional shock corresponding to the event; otherwise we do not change the process. We can continue with this procedure at time t and so forth because of the lack-of-memory property.

We also need to specify the effect of the generated jump event. It may well be a distribution itself. It may simply add a one-off extra charge to the process without changing the position X(t) that is needed to calculate X(t – 1) or change the position of X(t) itself, that is, shift the entire process. The specifics depend on the context.

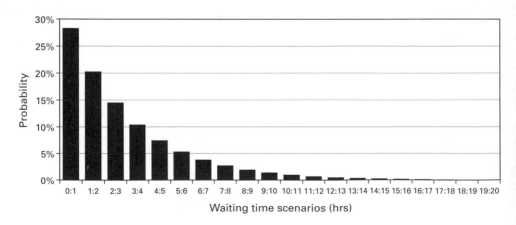

Figure E.9
Example exponential distribution.

Time Series Models of Higher Order

Random walk and mean reversion processes depend only on the last level of the variable X. This may make sense in some situations, but in others the path by which that last level was achieved can play a role in providing momentum. For example, if prices are dropping, one might argue that there is a tendency that they continue to go down. Such path dependence can be modeled with higher order models of the form:

$$X(t) = a_0 + a_1 X(t - 1) + \ldots + a_s X(t - s) + \varepsilon(t).$$

We can fit such so-called autoregressive models to the data just as random walk models. This leads into the vast domain of time series analysis, coverage of which goes well beyond the scope of this book. We refer the interested reader to the specialist literature.[9]

Appendix F: Financial Options Analysis

Options are commonplace in the world of finance. The theory of financial options is a well-developed academic field within the discipline of financial economics. Indeed, Robert Merton and Myron Scholes received the 1997 Nobel Prize in economics for their work with the late Fischer Black on this subject. Moreover, the use of financial options has grown rapidly since the 1990s and now constitutes a major part of the financial industry. The theory and practice of financial options have been highly successful.

Financial options are conceptually similar to options in an engineering system, to flexibility in design. They are rights but not obligations to future actions. In view of their similarities, and of their success in financial transactions, it is tempting to apply the theory of financial options to the valuation of options associated with technological projects and activities. These applications and theoretical extensions constitute the field of real options analysis.

However, the theory of financial options is of limited value in the context of the technology projects that are the focus of this book. We need to resist the temptation to apply the techniques of financial options blindly to such projects. We need to handle such applications with great care. This is because the context of technological projects differs significantly from that of financial transactions. The assumptions underlying the theory of financial options are not generally valid for projects, and that theory is thus of limited value for the design and implementation of technological systems. This appendix illustrates the fundamental ideas behind the theory of financial options by way of an illustrative example and then uses it to show why financial theory does not generally apply to projects.

The Fundamentals of Financial Options Analysis

Definitions

A financial option is a contract between individuals or institutions. One side, the "option writer," creates the option and then sells it to someone else, the "option holder." The contracts are for specified periods and define exactly when and how the option can be used. The two simplest financial options are the so-called "puts" and "calls."

A put option on a stock is a contract that gives the option holder the right but not the obligation to sell one share of the stock for an agreed price. A put option therefore acts as insurance against drops in the stock price. However, it is subtly different from traditional insurance on a house or a car:

1. Unlike the holder of car insurance, the holder of a put option on a stock does not necessarily ever have to own a share of the stock together with the put option. When the option holder decides to exercise the option, the contract is settled in a purely financial transaction: The option writer simply pays the difference between the agreed sales price and the market price of the stock on the day.

2. Unlike traditional insurance, the put option can be traded. The option holder can sell the option to someone else. The price of the option changes as the stock price changes. If the stock price drops, the chance increases that the insurance will be valuable in the future, and therefore the price of the put option increases.

The counterpart of a put option is a call option. This contract gives the option holder the right to *buy* a share of the stock for an agreed price at a specified time or at any time during a specified period. This call option is quite different from an insurance contract. It is more akin to a bet on rising share prices: The bet will pay off if the share price exceeds the agreed buying price; if not, the option is worthless.

Valuation of Options

To see how stock options are priced, consider a stock that is priced at $2.40 today and a call option on the stock that gives you the right but not the obligation to buy the stock at $2.50 in a week's time. For illustrative purposes, we assume that there are only two possible scenarios in a week's time: either an upside, on which the stock price rises to $3.00, or a downside that leads a drop to $2.00. Before we explain the pricing of the option, let's interpret today's share price of $2.40 in terms of this upside and downside uncertainty. In this simple world, the uncertain stock price can be decomposed into a sure payoff of $2.00, plus a gamble that will pay an additional $1.00 in the upside scenario and nothing in the downside. The market price for the certain payoff of $2.00 is of course $2.00 (assuming for simplicity that all figures are discounted to the present), and the market is prepared to pay $0.40 for a "stock gamble" that pays $1.00 in the upside and $0 in the downside. Figure F.1 illustrates the situation. Note that this description does not mention the probabilities of the upside and downside scenarios. We do not need them because there is already a market price for the uncertain payoff. The market factors in the traders' beliefs of the relative likelihood of these scenarios as well as their personal aversion to risk. There is no need to know the probabilities explicitly.[1]

Now consider a call option on the stock that gives you the right but not the obligation to buy the stock at $2.50 in a week's time. What's the value of this option? If the upside scenario occurs, you can buy a stock that is worth $3.00 for $2.50, so you make $0.50; if the downside scenario occurs, you could buy a stock worth $2.00 for $2.50—but of course you would not because you are not obliged. You would walk away without exercising your right, and the payoff from the option would be $0.

There are two ways to estimate the value of this option. One approach focuses on the probabilities of the outcomes. It estimates the probabilities of the upside and downside scenarios and then multiplies them with the respective scenario payoffs to get the expected payoff. If you estimate a 50 percent chance for the upside and downside scenarios, you obtain an expected value of $0.25 for the call option. Note that the market price for the stock, $2.40, is below the expected payoff of $2.50 because the market wants compensation for risk-taking. You may argue that the

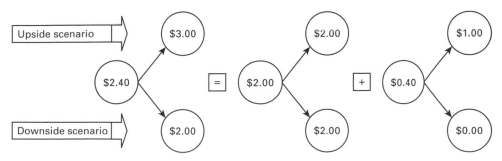

Figure F.1
Decomposition of share price into a sure payoff and a stock gamble.

market should also receive compensation for the risk in the option, and therefore the market price is likely to be lower than $0.25. But lower by how much? What is a sensible compensation for the risk in the option?

The alternative approach to valuing options exploits knowledge of the current stock price available in the market. Remarkably, it does not require explicit knowledge of the probabilities of the scenarios. It is clever and not at all intuitive if you are not an economist. Its development constituted a central insight into the creation of the theory of financial options. To appreciate how it works, recall from figure F.1 that the current stock price of $2.40 is the price for the sum of a sure payoff of $2.00 and a stock gamble that pays of $1.00 in the upside and $0 in the downside scenario. The market values the stock gamble at $0.40. Note further that in our case two of the call options provide precisely the same payoffs as the stock gamble (see figure F.2). Both deliver 2 × $0.50 = $1.00 if the upside scenario occurs and $0 otherwise.[2] Therefore, the two options should be worth precisely the same as the stock gamble, that is, $0.40. Based on this approach, the value of the call option is $0.20, not $0.25.

Which approach is "correct"? In general, each leads to a different value for the option. In the example, we obtained a value of $0.25, based on our estimates of the probabilities of the two scenarios, and a value of $0.20, based on the market price of the stock independent of probability estimates. Which should we choose?

The answer is that, in the marketplace, we must choose the value driven by the market prices. This is a practical matter, not a philosophical stance. The fact is that competition in the marketplace inexorably leads toward the value defined by the market price. To see this, consider the market's reaction to an offer to buy the call option for $0.25. Clever

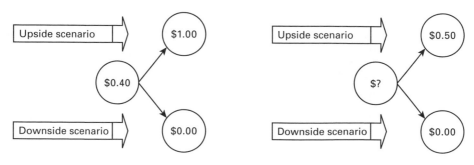

Figure F.2
Priced stock gamble (left) compared with call option that is not priced (right).

traders would realize that, because the uncertainty in the share price is the same as the uncertainty in two call options, this price provides the opportunity to make money with no risk. A trader would proceed as follows:

• Sell two call options for 2 × $0.25 = $0.50;

• Buy a share of the stock to counterbalance (or "hedge") the uncertainty;

• Finance this by borrowing $2.00 for 1 week from a bank, with the share as collateral. Because the share in the example has a sure payoff of $2.00, the loan would be riskless for the bank, and therefore the interest rate would be low.

• Now the trader has $2.50; $2.00 borrowed from the bank and $0.50 received from the sale of the call options. He has bought a share of the stock, which cost $2.40, and he will pocket the remaining $0.10 as riskless profit.

Why is the profit of $0.10 riskless? To see this, look at the possible scenarios in a week's time. If the upside scenario occurs, the clever trader will sell the share for $3.00 and use this income to repay his loan of $2.00 (assuming the interest was negligible) and to pay 2 × $0.50 to the option holder, who will of course exercise the two call options. If the downside scenario occurs, the trader will sell the share for $2.00 and repay the loan. In neither case does he have to tap into his pocket with the $0.10 profit. His profit is riskless. The technical term for such riskless profits through clever trading is "arbitrage." Here we can effect the arbitrage trading by replicating the payoff of the option through a combination of a loan and an investment in the stock (known as a "replicating portfolio"). The traders' eagerness to make money easily by buying the options drives down the price of the option until there is no riskless profit available, that is, until the price of the option reaches the level determined by the market-based approach. This process is "arbitrage-enforced pricing." This phenomenon determines the value of options in the marketplace.

This example illustrates the fundamental assumptions at the heart of the Nobel Prize-winning theory and procedures for valuing financial options:

1. It is possible to replicate the uncertain future payoffs of an option through a combination of already traded assets. This replicating portfolio has precisely the same payoffs as the option in all possible future states of the world.

2. Because efficient markets do not allow for arbitrage gains, that is, for returns without risks, the correct price for the option will equal the known price of the portfolio of traded assets that replicates the payoffs of the instrument in each future scenario.

Inapplicability of Financial Options to Design

Both financial and design options provide the right but not the obligation to a future action. It is therefore tempting to try and apply the beautiful theory of financial options to the valuation of design options. However, the application of financial options theory to real projects is problematic. This is because critical assumptions behind financial options pricing generally do not apply to engineering systems and technological projects.

The key assumption is that it is possible to create a portfolio of traded assets that will replicate the payoffs from the option in every possible future state of the world. To apply financial options pricing theory, we therefore have to answer two critical questions:

1. What is the replicating portfolio?

2. Where can we buy and sell the constituents of this replicating portfolio?[3]

In the case of financial options, the answer to the second question is obvious: the financial markets—the places where we can buy and sell financial assets. The first question, however, is hard to answer even for financial instruments. In fact, the work of Black, Scholes, and Merton created the new discipline of financial engineering, which is largely concerned with this challenge.

In the case of engineering systems, in our experience, it is impossible to answer the above questions satisfactorily for most real projects.

A first difficulty lies in trying to define an appropriate replicating portfolio. For example, what would it be for the option to enlarge a motorway from four to six lanes? The value of this option is driven by demand for traffic on this specific road. Conceptually, the option to extend the motorway from four to six lanes is a call option on demand. If demand is high, it makes sense to exercise the option; otherwise the option remains dormant. The replicating portfolio for this call option would, as illustrated above, consist of a combination of borrowed risk-free money and buying or selling shares in the underlying stock that the option is written on. However, unlike common stock, demand for traffic between two cities is not a traded commodity. To apply financial options pricing, we would therefore need a traded financial surrogate that goes hand in hand with demand for car travel on that particular road.

Second, we would have to make sure that this financial surrogate remains a good proxy for financial demand over the duration of the option. Unlike financial options, options on projects are likely to be long term and may not even have fixed expiration dates. For example, the widening of a motorway might not be considered for 5 or 10 years. By contrast, financial options will have a clearly defined expiration date, typically within weeks or months of the first issue of the options contract. Their relatively short durations give some credibility to the idea that recent historical data can represent the distribution of future returns of the assets in a replicating portfolio, an assumption we need for calibrating financial options models. In fact, although the assumption that market data represent future returns may be credible in a benign economical climate, it is much less so when the market changes fundamentally, as in the 2008 financial crisis. The assumption is even less credible in relation to engineering design options, which will typically have long time horizons of several years, sometimes decades. For example, Portugal added the second deck to its 25 April Tagus Bridge about 20 years after initial

Table F.1
Context differences between financial and design options, which make it inappropriate to apply theory of financial options to flexibility in engineering systems

Financial option	Design option
Asset is widely replicated (company stock, commodities, financial assets)	Asset is unique (a bridge, a building, a new product)
There is a market for such assets	No market, in general (may be for product, such as for copper from mine)
A replicating portfolio can be created	Unlikely to be able to create replicating portfolio (maybe in short term, as for traded commodities such as copper, but unlikely over years of option)
Option valid for months	Option for years, even decades
Market data available	Data spotty, maybe unavailable
Recent market data credible to anticipate market variations over life of option	Historical data do not anticipate trend-breakers likely to occur over long life of option
Option characteristics well defined (strike price, time to maturity, payoffs)	Option characteristics may be unclear and change over time (indefinite life of option, indefinite exercise type, size, and price)

construction, and the HCSC doubled the height of its Chicago skyscraper after a decade (see box 1.6).

Third, design options may not have clear, predefined conditions for their exercise or payoff structure. This contrasts with financial options that we can exercise when the stock price is appropriately high or low. Furthermore, the exercise of design options changes features or characteristics of the system, which is not the case for financial options. An expansion of an airport, motorway, or hospital may well in itself lead to a demand push. Overall, the assumptions at the base of financial options rarely apply to the design of technological systems and products, as table F.1 summarizes.

In summary, although financial options pricing is a conceptually appealing theory, it is typically not suited for a valuation of options in the kind of technology projects dealt with in this book.[4]

Notes

Chapter 1

1. de Weck et al. (2004) discuss in detail how Iridium could have managed its satellite deployment better.

2. Babajide et al. (2009) discuss the standard design for oil field development and its alternatives.

3. Gessner and Jardim (1998) provide technical details on this double-decking process.

4. Hassan et al. (2005) describe the analysis for the satellite example.

5. Guma et al. (2009) describe the case. Guma (2008) and Wittels and Pearson (2008) provide more details.

Chapter 2

1. Savage coined the "flaw of averages." He describes the concept in detail in Savage (2009). Yang (2009) provides an extended example for the case of capacity planning in the auto industry. Appendix A discusses the details.

2. Several comprehensive or meta-studies of the inaccuracy of forecasts are available. Ascher (1978) provides a seminal text. Miller and Lessard (2001) give extensive examples of the inability of planners and designers to estimate uncertainties accurately.

3. See Blix et al. (2001).

4. The data on variations between forecast and final results come from the extensive work that Flyvbjerg and his team have presented in many places. See in particular Flyvbjerg et al. (2003, 2005).

5. Data on commodity prices are available from many sources on the web. Figures 2.5 and 2.6 are adapted from kitco.com and <http://mining101.blogspot.com/2008/06/rising-copper-prices-copper-basics-bhp.html>.

6. Data on changes in perceived estimates of oil reserves come from Lin (2009). See also Babajide et al. (2009).

7. Many descriptions of the evolution and use of UAVs are available. Much of the attention focuses on the largest versions, the Global Hawk and the Predator.

Chapter 3

1. The case paraphrases an actual project in the Bluewater development in England (<http://www.bluewater.co.uk>). The numbers used are representative but do not correspond to those of the actual project. Chapter 6 and appendix D also discuss this garage case. A version of it appears in de Neufville, Scholtes, and Wang (2005).

2. Several open source and commercial software packages are available to perform these calculations. We used the commercial package XLSIM® to calculate and graph the results in this book.

3. A histogram divides the range of a distribution into subdivisions of equal size, counts the number of occurrences of an outcome within each smaller range, and shows this number as a vertical bar. This process shows the frequency of ranges of outcomes, which is the probability distribution. In the example, the Monte Carlo software took the range of the 10,000 realized NPVs (approximately between –$35 million and +$11 million) and divided it into 10 regions of equal width, about $4.6 million in this case. The height of the bar associated with each bin measures the fraction of the 10,000 realized NPVs that fall into the respective range.

4. The concept of "value at risk"(VAR) has been widely used in finance. It refers to the amount that might be lost or the target that the design might not attain, with a specified probability. It generally specifies a time horizon for the risk. The "value at gain" (VAG) is the complementary probability of what might be gained. We sometimes refer to the cumulative probability curve as the "value-at-risk-and-gain" (VARG) curve.

5. The economies that designers can achieve by building capacity all at once include both economies of scale and avoided costs of interrupting operations as facilities are expanded. Unless designers plan future expansion carefully, it can be problematic to construct around ongoing operations. For economies of scale, see appendix C.

6. See Cardin (2007) for a discussion of the effect of many different decision rules for expansion.

Chapter 4

1. See, for example, Shishko et al. (1995). Their *NASA Systems Engineering Handbook* sets out the accepted practice, which begins with defining the requirements. Similarly, architects typically work to architecture programs that define the number and types of spaces they have to provide for a facility.

2. See chapter 2, especially the section on "The standard forecast is 'always wrong'."

3. See the seminal text by Manne (1967). As presented in appendix C on the *Economics of Phasing*, his approach leads to interesting insights. However, his implicit concept of overall demand as the appropriate system driver is overly simplistic.

4. For example, countries define socially important measures politically. To count officially as "unemployed" in the United States, a person has to have been employed previously and to be actively looking for a job. The French tend to focus on whether you have a job, even if you have given up looking for one. In Japan, the government pays companies to keep people salaried even if there is no work for them; this is functionally equivalent to an unemployment dole, but Japan does not classify the recipients as unemployed (see Innes, 1975).

5. R-squared is also known as the coefficient of determination. Technically, 100 times R-squared is the percentage of the variance in the data that can be attributed to the estimated trend. If R-squared is 0.70, the trend captures 70 percent of the variability in the data, and the remaining 30 percent of the data variability remains "unexplained."

6. Scenario analysis forms the basis of strategic military planning. Its use in business goes back to the oil crises of the 1970s. Subsequently, petroleum companies, most notably Shell, developed and popularized this planning tool for civilian use (see Lingren and Bandhold, 2009). Schoemaker (1991) provides a useful description of how scenario planning can be organized for a group.

7. Statistical analyses of historical data series can show excellent statistical fit, with high R-squared values that can be deceptive and lead unthinking users to believe that predictions from such analyses are particularly accurate. This is not the case because good fit does not imply predictive power. There are two main reasons for a good statistical fit of statistical

models that involve time. The first is that any factor that changes exponentially (i.e., at a given growth rate) correlates well with any other exponentially changing factor. Consequently, it is easy to find good candidates for correlation. Second, it is a mathematical fact that the overall correlation can always be improved by adding more variables to the equation. Thus, forecasters do not generally find it difficult to obtain good statistical metrics if they want to do so.

8. The case reflects a real-life situation. We changed the name of the hospital and added a small amount of random noise to the data. A teaching case of Lee and Scholtes (2010) provides more details.

9. See, for example, Ulrich (2003) and Marberry (2006).

10. Exponential smoothing is a particularly prevalent forecasting procedure in practice. Comprehensive reviews can be found in Gardner (1985, 2008).

11. The model anchored at the most recent data point is called a random walk model. Mean reversion models offer a compromise between trend line models, where the best estimate for next year is the point on the line, and random walk models, where the best estimate for next year is anchored at the most recent observation. Appendix E, Dynamic Forecasting, gives more details.

12. In a more realistic model, we would organize the population of fertile women into segments with different fertility rates. For example, foreign-born women may have a significantly different fertility rate. This matters if the proportion of foreign-born women is likely to change.

13. See Tromans et al. (2009).

14. The traditional view is available in the publications of the International Air Transport Association (IATA) and the U.S. Federal Aviation Administration (FAA).

15. There are many examples of airlines disappearing and changing the need for airport facilities: The collapse of Swissair and TWA left Zurich and St. Louis airports with large empty terminals. Similarly, the reorganizations of U.S. Airways and Alitalia caused these airlines to abandon their hub operations at Pittsburgh and Milan.

16. It is generally not economical to design for absolute peaks in demand. The facilities might be unused for the rest of the year. Usual practice is to design for some fraction of absolute peak demands and to put up with crowding, delays, or otherwise poor service during absolute peaks. Designers thus make trade-offs. What is "reasonable" depends on the circumstances.

17. de Neufville and Odoni (2003) and de Neufville and Belin (2002) provide detailed explanations.

18. See Nababan (1993) for a case study of Boston's experience.

19. For details, see Maseda (2008).

Chapter 5

1. The original quote is: "It can scarcely be denied that the supreme goal of all theory is to make the irreducible basic elements as simple and as few as possible without having to surrender the adequate representation of a single datum of experience" (Einstein, 1933).

2. Screening models have a long history in the design of projects. Jacoby and Loucks (1972) (available in de Neufville and Marks, 1974) published one of the first major applications. They used simple linear models of the dynamics of river basins to identify plausible development strategies for the Delaware River. They then investigated those possibilities with detailed stochastic models of the hydrologic flows.

3. The number of development paths increases exponentially with the number of periods because system operators react to prices and change the system as prices change. This reality contrasts with the use of binomial expansions in the financial analysis of options, according to which the total number of end states equals the number of periods plus one. The use of the binomial expansion assumes "path independence," that is, that where you

end up is independent of how you get there. Path independence can make sense for prices. Path independence is not an appropriate assumption to describe a system: The future of a project generally depends very much on the previous history, not just its current state. For example, if demand has been low for a long time and has made a sudden jump to today's level, managers are less likely to expand capacity than when demand has been moderately but continuously increasing over time to today's level, and managers could mentally prepare for an expansion. Today's state, the level of demand, may be the same in both situations, but the path taken to get there was different and therefore the decision can be quite different.

4. Gigenrenzer (2002) provides an excellent account of these issues.

5. The lack of acceptance of financial options analysis among engineering designers is an example of how people often distrust complex models. Black–Scholes and similar financial processes for evaluating the value of options are indeed difficult to understand. When presented with the results of these models, decision makers find themselves being told what to do by junior staff based on impenetrable assumptions. Moreover, the assumptions used by financial analysts—essential to the operation of their analyses—are often inappropriate. For example, the financial analysis of options normally assumes that future uncertainties derive from "stationary" processes whose mean and variance do not change over time. In reality, this assumption is not valid over long periods, over the life of interesting systems in particular (see appendix F, Financial Options).

6. Note that we do not talk about the best solution. This is because the model is a simplification, and our assumptions about the distribution of the uncertainties can hardly ever be fully justified through external evidence. Moreover, as chapter 6 explains, the notion of what is "best" is clear only in the simplest cases. In general, any design has many qualities that we cannot legitimately reduce to a single scale. Furthermore, the different stakeholders in a system will price attributes differently and prefer different solutions.

7. In most models, every variable will be in some way related to (at least a fixed proportion of) every other variable in the model. So the number of computational relationships is roughly (a fixed proportion of) the square of the number of variables in the model.

8. Mathematical jargon refers to simulation models as "response surface models." The name derives from the idea that the value of a system is a function of the design parameters and that this value function is a surface in n-dimensional hyperspace. This explains why we use an alternative descriptor.

9. Forrester (1961) originally elaborated systems dynamics models. His colleagues and students at MIT and elsewhere have improved the techniques and applied them to many areas. The essence of the approach consists of chains of actions or events that both influence downstream events and have feedback loops that can either dampen or amplify upstream events. A range of software exists for the implementation of systems dynamics models, such as Stella®, Powersim®, and Vensim®.

10. Optimization procedures generally fall into one of three types: linear programming and its derivatives; dynamic programming; and heuristic search methods, such as genetic algorithms or simulated annealing. Linear programming methods exploit mathematical features of the problem—specifically the convexity of the feasible region—to compute an optimal solution, if it exists. They are fast and reliable even for very large problems, but they have a limited application range due to their restrictive mathematical assumptions. Dynamic programming is particularly relevant when the system to be optimized consists of sequences of independent decisions, most obviously when it evolves over time. Linear and dynamic programming methods are very powerful for the specific cases for which they are applicable. Heuristic search methods are less powerful but more versatile. These apply generic search strategies to detect regions in the solution space that are likely to contain a good solution and then zoom in toward a possible optimum. Heuristic methods are easily implemented and can be adapted to a wide range of optimization problems without much concern for their mathematical properties. Their downside is that, as a rule, they cannot guarantee optimality and can be quite slow in high-dimensional problems.

11. Suh (2005) provides a detailed example of flexibility in automobile production based on platform design.

12. See Frey and Wang (2006).

13. Yang (2009) provides full details.

14. See Brushlinsky et al. (2006).

15. See, for example, Routledge (1980), Sneeuwjagt (1998), Thompson et al. (2000), Tolhurst et al. (2006), Brillinger et al. (2004), and González et al. (2005).

16. Wang (2005) describes this case in detail.

17. Lin (2009) discusses this case in detail.

18. See Steel (2008).

19. Lin (2009) provides details on this case and the great increase in value.

Chapter 6

1. Crystal Ball®, @Risk®, and RiskSolver® are widely used commercial software packages. Savage's XLSim® software, used to produce the graphics in this book, is a good entry-level package, considerably less expensive but sufficiently powerful for many practical applications. As standard in the field, each allows users to choose the kind of distribution they want to use, including uniform, normal, triangular, lognormal, and many others. Each also enables the analyst to prepare attractive graphical representations. Other simple add-ins to Excel® spreadsheets are also freely available from the web. As expected, the free add-in modules are not as sophisticated or user-friendly as the commercial products.

2. The question of how many evaluations a Monte Carlo simulation should run has no obvious answer. If we could be confident in the description of our input uncertainties, we could in principle estimate the runs needed to develop a specified level of confidence in the accuracy of the simulation. However, we are not likely to be confident in the long-range forecasts of important factors. Therefore, estimations of run lengths are unlikely to be of much use. Because we can run each simulation quickly, it is good to use a generous number of samples. This reasoning explains the practice of using 1,000 or more simulations. If the model is complex and computational resources are limited, however, one may have to settle for fewer. Monte Carlo simulation software will normally report standard errors and confidence intervals for estimated statistics, which become narrower with an increasing number of simulations and give a guide to the accuracy achieved.

3. As a semantic caution, note carefully that this value is "expected," in the mathematical sense that it is the average value. It is not necessarily the most likely value. The expected value may even never happen. For example, the average value when rolling a six-sided die is $3.5 = (1 + 2 + 3 + 4 + 5 + 6) / 6$, a number that of course never shows up.

4. A skew in the distribution refers to the fact that one tail of its shape is longer than the other. The tail that is longer, say the lower 50 percent of the distribution, pulls the average toward that side. The median, however, will stay put if there is no change in the other tail, say the upper 50 percent. Thus, the difference between the median and the mean is an indication of skewness.

5. In financial markets, an option is a contract between traders that gives the owner the "right, but not the obligation" to take an agreed future action at agreed terms. A "call" option gives the owner the right to buy something valuable at an agreed price if the right opportunity presents (such as a higher market price for the asset). A "put" option gives the owner the right to get rid of a property or asset at an agreed price if the market price of the property dips sufficiently. This helps to avoid losses. An insurance contract is a form of put option. For example, if you total your car and your insurer pays you for it, you have essentially sold this now worthless property for a good price.

6. See, for example, Kulatilaka (1993).

7. This is a bit subtle. To see that this is the case, suppose the performance of two designs A and B is determined by rolling a die and that design A achieves an economic value of

100 when the die shows 1, 2, 3, 4, or 5 and a value of –100 when the die shows a 6, whereas design B yields –200 if the die shows 1, 2, 3, 4, or 5 and 50 when the die shows a 6. Design A's target curve is entirely to the right of design B's, hence design A's chance of missing any given target is smaller than design B's. However, if the die shows a 6, design B outperforms design A.

8. A utility function embodies an overall expression of value for all dimensions of choice. Theoretically, we could apply utility functions to translate multiple dimensions of choice into a single metric, and we could thus avoid the problem of choosing among multiple dimensions. However, this is not practical. Reliable, accurate measurement of the utility function for any individual requires careful extended laboratory procedures. In practice, we cannot apply these methods to decision makers, who have neither the time nor the appetite to be experimental subjects. Despite extended efforts to apply utility theory to the evaluation and selection of major projects, this approach has not worked out to be practical.

Simple textbook descriptions of how to measure utility in rapid interviews led to inconsistent results. de Neufville and McCord (1986) describe the kind of controlled analyses required to develop meaningful results. Starting with the seminal analysis of plans for a new airport for Mexico City (Keeney and de Neufville, 1973), there have been many attempts to promote utility analysis as a means of dealing with multidimensional decision making (e.g., Keeney and Raiffa, 1993), but these have not proved practical for the evaluation and selection of major projects. Finally, note that it is impossible, from a theoretical perspective, to define the utility function of the group of decision makers without specifying the exact voting rules that will prevail among them, as Arrow (1950) demonstrated. In short, attempts to use utility functions to define the best choices for a system are futile from a practical perspective.

In this context, analysts should be particularly careful about the simplistic approach that defines the utility function as the linear sum of the metric for each dimension times a relative value or weight for each metric. This approach is appealing insofar as it would dodge any complicated judgments. However, this linear approach is fundamentally misguided. Theory indicates that utility functions are nonlinear according to the principle of diminishing marginal utility—that is, the relative weight given to any factor depends on one's level of satisfaction. For example, if you have no food, you might pay a lot for a meal; but if you already have a lot of food, you might not value extra food highly at all. Experiments confirm this general rule. Thus, any scheme for defining the overall best design based on relative fixed weights is highly suspect.

9. Concern about regret can be thought of as a form of risk aversion, of wanting to avoid decisions that might look wrong in retrospect, of preferring "to be safe than sorry." Theoretical discussions define the measure of regret as the difference, for any set of circumstances, between what actually results from a choice and what the chooser could have obtained from the alternative most suitable to those circumstances.

10. In design, it is conventional to define "robustness" in terms of a ratio, the coefficient of variation, which is the standard deviation of the performance compared to its average value. A design with the lowest ratio is said to be the most robust. Robustness is a desirable characteristic of design in many cases, particularly when it is possible to define a single most desirable value, such as when one wishes to tune into a broadcast at a fixed frequency.

However, robustness is not desirable when the goal is to maximize (or minimize) a value, for example, when we want to maximize profit. In cases where we want to maximize value, we would like to minimize the downside and maximize the possible upside opportunities. In terms of flexible designs, we want solutions with skewed value distributions, with limited downside and great upside potential—the greater the better—that is, a design with higher standard deviation is more desirable than one that limits upside gains. This point is counterintuitive to those who think that more robust is always better; it certainly is not.

11. Petroleum engineers distinguish between the total oil in a reservoir and the amount they can recover economically. The former is known as the Stock Tank Oil Initially in Place

(STOIIP). The reserves, that is, the amount that it is economically worthwhile to extract, are some fraction of the STOIIP, maybe between 1/3 and 1/2.

12. It is common for management in extractive industries (mining of all forms in addition to oil and gas) to define a long-term price of the extracted product that designers should use in valuing possible projects. They do this to create a level playing field among the design teams working on different projects across the company. Remarkably, management frequently changes these statements about long-term prices often within a few years. The prices actually used by a company are often highly confidential as they reveal how aggressive a company might be in bidding for reserves.

13. For a good example of how it is possible to explore trade-offs in system design, see the work by Cohon and Marks (1974) on the development of river flows for agricultural, industrial, and other purposes, available also in de Neufville and Marks (1974).

Chapter 7

1. Large Canadian airports, such as Calgary, Edmonton, Toronto, and Vancouver, have designed their terminals with moveable partitions to enable them to direct passengers to the segregated international facilities or not depending on need. Other leading airports do likewise, as de Neufville and Odoni (2003) describe. Progressive manufacturers of major equipment, such as cars or tractors, can similarly switch their production rapidly by changing the jigs or redirecting the robots on their assembly lines.

2. For details, see Hassan et al. (2005) on satellite fleets and Lin (2009) on oil platforms.

3. This is a common issue. From the point of view of final design, it is desirable to have the climate control plant as close as possible to the main buildings to reduce the cost of conduits and cut energy losses. However, it is almost impossible to move such plants because they are vital to the operation of the buildings and have to keep operating 24/7. The Dartmouth-Hitchcock Medical Center offers a good example of how the placement of their heating and chilling plant enhanced the possibility of future expansion.

4. See Guma et al. (2009) and Wittels and Pearson (2008) for full descriptions.

5. See Barkley (2006) and American Institute of Architects (2007, 2010). Different versions of the process exist, and some people refer to integrated project management.

6. This refers to the so-called South Suburban Airport near Peatone, Illinois. Its future is currently doubtful because the City of Chicago has focused attention on rebuilding its O'Hare airport. The U.S. FAA has primarily worked with the State of Illinois. Airport authorities elsewhere have also acquired land for possible future airports in advance of a possible decision to build, such as Sydney, Toronto, and Bangkok—which eventually did build its Suvarnabhumi airport on such a site.

7. Wittels and Pearson (2008) discuss the HCSC and the Tufts cases in detail.

8. Walker (2001) has discussed the concept of signposts.

9. de Neufville et al. (2008) provide detailed discussions of hospital developments.

10. McConnell (2007) gives interesting accounts of some of the politics surrounding the Houston metro.

Appendix A

1. See Savage (2009).

2. See Shisko et al. (1995).

3. When base-case assumptions are replaced by distributional assumptions (e.g., on the future price of oil), it is critically important that all projects being evaluated use the same distributions for common inputs. Technology is available to facilitate probabilistic planning across projects (see Savage et al., 2006; http://www.probabilitymanagement.org).

4. The engineering curricula at leading European and Asian universities commonly place little emphasis on economic or social studies. In the United States, the accreditation boards

require engineering undergraduates to take a quarter of their credits in some form of liberal arts. However, the professional curriculum often does not integrate this material.

5. To be precise, Jensen's law is somewhat more specific. It states that $f(E[X]) \leq E[f(X)]$ if $f(.)$ is a convex function.

Appendix B

1. The argument against using DCF analysis when cash flows are uncertain centers on the inability either to account for optionality, such as the option to defer investments, or to vary the discount rate as risk varies as the project advances. See Trigeorgis (1996), for example. From a theoretical perspective, the discount rate should reflect the degree of risk: the greater the uncertainty, the greater the discount rate. However, greater uncertainty in the cash flows gives value to the option to wait with the investment until the uncertainty is resolved to some degree. The time value of money competes with the value of learning about the future, and the latter is not included in DCF analysis. Thus, finance theory has developed what is known as "options analysis." This uses a "risk-neutral analysis" that transforms the distribution of future cash flows into an alternative "risk-neutral" distribution, which in turn justifies a constant discount rate, in fact a "risk-free rate." Theory justifies this approach on the basis that broad markets will enable "arbitrage enforced pricing" of the assets generating the cash flows. Appendix F provides details.

Financial markets have widely implemented options analysis. The functioning of these markets most often justifies the assumptions under which Merton, and Black and Scholes (1973) pioneered this approach. However, these assumptions generally do not apply to real-world projects we might design. Moreover, options analysis requires special training. In any case, many practicing designers and system managers have rejected the use of real options based on finance theory. They neither understand nor trust it.

2. In the United States, the Office of Management and Budget sets the discount rate for agencies of the national government. Its current pronouncements appear in Circular A-94, for example, U.S. Office of Management and Budget (2008).

3. See Brealey and Myers (2002) for details.

4. At a deeper level, the discount rate reflects available opportunities. This is because organizations and investors will rationally select the projects with the highest returns, and these will set the bar for the price of the capital invested.

5. See Brealey and Myers (2002).

Appendix C

1. Fixed setup costs and project overheads are a second common cause of economies of scale. Some costs, such as those of a project management team, do not scale with the size of an enterprise. An organization has to have them whether the project is small or large. The formula is then Total Cost = Fixed Cost + K*Capacity, and therefore Average Cost = Fixed Cost / Capacity + K: the larger the capacity, the lower the effect of fixed costs per unit of capacity. The formula is different, but the effect is the same as indicated in the text.

Appendix D

1. A student version of XLSim® is included in Sam Savage (2004), which is an excellent tutorial to Monte Carlo Simulation.

2. The tornado chart in figure D.3 was produced with the Sensitivity Toolkit by Baker, Powell, and Burnham (see <http://mba.tuck.dartmouth.edu/toolkit/index.html>)

3. We can use two-way data tables to produce sensitivity charts that show how system performance changes when two inputs vary simultaneously. However, associated three-dimensional graphs are somewhat difficult to interpret and are limited to a maximum of two variables.

4. Figure D.4 is part of a data table with 1,000 trials, stored in the range T10:AB1010 within the randomized NPV spreadsheet (i.e., new inputs and an associated NPV were sampled every time F9 is hit). The data table was produced by (i) filling column T11:T1010 with consecutive numbers 1–1,000, (ii) linking U10:AB10 to the cells that contain the generated inputs and outputs that we wished to track, (iii) highlighting T10:AB1010, (iv) invoking the data table command, and (v) nominating an arbitrary unused cell, e.g., S10, as "column input cell." Note that the data table command does not work across worksheets, so the changing inputs and outputs have to be in the same sheet as the data table.

5. The statistically trained reader will realize that the theoretical correlation between demand and cost in this example is precisely zero.

6. The open source statistics package R has a function to create a scatter plot matrix. SAS sells a more user-friendly commercial package with many more graphic choices under the name of Jmp.

7. The formula "=0.5*(RAND() + RAND())" generates a triangular distribution between 0 and 1. The formula "=A1 + 0.5*(RAND() + RAND())*(B1 – A1)" generates a triangular distribution between a number in cell A1 and a number in cell B1 with a peak at $(B1 - A1)/2$.

8. If R is the annual revenue when independently sampled as in step 4, R' the new revenue variable, and D the demand deviation, then the formula is $R' = R + b*D$. Therefore $E[R'] = E[R + b*D] = E[R] + E[b*D] = E[R] + E[b]*E[D] = E[R]$. Here we have used the fact that $E[X*Y] = E[X]*E[Y]$ when X and Y are independent and that the expected deviation of demand from its projection $E[D] = 0$.

9. If the shocks can be large, it may be necessary to safeguard against negative demands, for example, by using a formula of the form $X(t) = max\{0, X(t - 1) + \varepsilon(t)\}$. This leads to a slight violation of the expectation consistency requirement, as discussed at the end of appendix D. However, the bias is typically small and acceptable in most practical situations.

10. This is the inverse transform method. To see why it works, recall that Prob(RAND() ≤ z) = z for all z with $0 \leq z \leq 1$. Hence, Prob(X ≤ x) = $F(x)$ = Prob(RAND() ≤ F(x)) = Prob(F^{-1}(RAND()) ≤ x) and therefore X and F^{-1}(RAND()) have the same distribution.

11. Analysts often use exponential distributions (or, more generally, a so-called Gamma distribution) to model uncertain durations, such as the time between two events. The exponential distribution with mean m has the cumulative distribution function $F(x) = 1 - exp(-x/m)$. Hence, its inverse is $F^{-1}(u) = - Ln(1 - u)*m$. Because 1-RAND() has the same distribution as RAND(), the formula in the text generates an exponential distribution with mean in cell A1.

12. This result is the central limit theorem. It essentially says that the distribution of the sum of many well-behaved, unrelated, uncertain numbers tends toward a normal distribution no matter what the distribution of the summands might be (see e.g., W. Feller, 1968).

Appendix E

1. Appendix E uses fixed nonrandom rates. In general, they may well be time dependent, that is, the model may be of the form $X(t) = a(t) X(t - 1) + \varepsilon(t)$.

2. After n = 3 moves, there are $2^3 = 8$ possible states, achieved by the following eight move sequences (u = up, d = down): uuu, uud, udu, duu, udd, dud, ddu, and ddd. In the recombinant situation, the moves uud, udu, and duu lead to the same end-point, and so do udd, dud, and ddu, which leaves us with only n + 1 = 4 end-states.

3. See, for example, Lee and Scholtes (2010) for more detail.

4. See Fildes and Makridakis (1995) for more detail.

5. Microsoft Excel has functions SLOPE, INTERCEPT, and STEYX to calculate the slope and intercept of the model, as well as the standard deviation of the residual of a regression of one array of data onto another.

6. A Kolmogorov–Smirnov test would be an appropriate way to test this statistically.

7. The exponential distribution is a special case of a Gamma distribution, which is more versatile. Because of the monotonicity of the histogram of the exponential distribution (see figure E.9), the next event is more likely to happen earlier than later, most likely in the immediate future. Gamma distributions allow you to model situations where the next event may be more likely to happen in the more distant than in the immediate future.

8. Take the duration until a fish bites as an example. Suppose you have been trying to catch a fish for half hour to no avail, when your friend arrives with his gear. He starts a half hour late, but his chance of catching the first fish are now the same as yours. The fact that you didn't catch a fish in the past half hour does not improve your chance of catching a fish over the next half hour.

9. See, for example, Shumway and Stoffer (2006) or Hamilton (1994).

Appendix F

1. Sharpe (1993) makes a persuasive argument of a price interpretation of probabilities in markets.

2. In general, it is possible to establish a relationship between the payouts of the call option, the stock price, and a risk-free asset such as a secure government bond under quite general assumptions, which are typically taken to be satisfied in major financial markets. This will establish the value of the option through a combination of already traded assets and lead to the celebrated Black–Scholes formula but is neither obvious nor worth explaining for our purposes. Luenberger (1996) provides a good explanation of the process. The example here bypasses this mathematical effort, which gives no further fundamental economic insight.

3. A third question is less obvious but equally important. Is it possible to trade the replicating portfolio with little transaction cost? This question is relevant because the theory driving the pricing of financial options assumes that traders in options can continuously "rebalance" the replicating portfolio, that is, sell parts of some assets and buy more of others. This is because variations in the price of a stock influence the price of the option and thus change the desired composition of the replicating portfolio. For example, increases in the price of a stock increase the value of a call option for that stock and the likelihood that its owner will exercise it. It is then not enough to adjust the market prices of the assets in the existing replicating portfolio; it is necessary to rebalance the entire portfolio to account for the new information. For this to be practical, it must be affordable. This means that one needs a fairly "frictionless" market, one with low transaction costs.

4. Borison (2003, 2005) gives a more elaborate critique of the application of financial options analysis to real options valuations.

Bibliography

American Institute of Architects. *Integrated Project Delivery: A Guide.* Washington, DC. 2007.

American Institute of Architects. *Integrated Project Delivery: Case Studies.* Washington, DC. 2010.

Argote, L., S. Beckman, and D. Epple. "The persistence and transfer of learning in industrial settings." *Management Science* 41 (11) (1990): 140–154.

Arrow, K. "A difficulty in the concept of social welfare." *Journal of Political Economy* 58 (4) (1950): 328–346. http://en.wikipedia.org/wiki/Arrow%27s_impossibility_theorem - cite_ref-0

Ascher, W. *Forecasting: An Appraisal for Policy-Makers and Planners.* Baltimore, MD: Johns Hopkins University Press, 1978.

Babajide, A., R. de Neufville and M.-A. Cardin. "Integrated method for designing valuable flexibility in oil development projects." *Society of Petroleum Engineers, Journal of Projects, Facilities and Construction* 4 (2) (2009): 3–12.

Barkley, B., Sr. *Integrated Project Management.* New York: McGraw-Hill, 2006.

Black, F., and M. Scholes. "The pricing of options and corporate liabilities." *Journal of Political Economy* 81 (1973): 637–659.

Blix, M., J. Wadefjord, U. Wienecke, and M. Ådahl. "How good is the forecasting performance of major institutions?" *Economic Review* 3 (2001): 38–68. http://www.riksbank.se/upload/Dokument_riksbank/Kat_publicerat/Artiklar_PV/er01_3_artikel3.pdf

Borison, A. "Real options analysis: where are the emperor's new clothes?" *Journal of Applied Corporate Finance* 17 (2) (2005): 17–31. 2003 version available at http://www.cob.sjsu.edu/webb_k/B260/Black-Scholes.pdf

Brealey, R., and S. Myers. *Principles of Corporate Finance.* 7th ed. New York: McGraw-Hill, 2002.

Brillinger, D., R. Burgan, and J. Benoit. "Probability based models for estimation of wildfire risk." *International Journal of Wildland Fire* 13 (2004): 133–142.

Brushlinsky, N., S. Sokolov, P. Wagner, and J. Hall. *World Fire Statistics.* Paris: CTIF (Comité Technique International de Prevention et d'Extinction Du Feu), Centre of Fire Statistics, 2006. http://ec.europa.eu/consumers/cons_safe/presentations/21-02/ctif.pdf

Cardin, M.-A. *Facing Reality: Design and Management of Flexible Engineering Systems.* Cambridge, MA: S. M. Thesis, MIT Technology and Policy, 2007.

Cohon, J., and D. Marks. "Multiobjective analysis in water resource planning." *Water Resources Research* 9 (4) (1973): 333–340 (and chapter 21 in de Neufville and Marks, 1974).

de Neufville, R. *Applied Systems Analysis: Engineering Planning and Technology Management*. New York: McGraw-Hill, 1990. http://ardent.mit.edu/real_options/ASA_Text//asa_ch4.pdf

de Neufville, R., and S. Belin. "Airport passenger buildings: Efficiency through shared use of facilities." *American Society of Civil Engineers Journal of Transportation Engineering* 128 (3) (2002): 201–210.

de Neufville. R., and D. Marks, eds. *Systems Planning and Design: Case Studies in Modeling, Optimization, and Evaluation*. Englewood Cliffs, NJ: Prentice-Hall, 1974.

de Neufville, R., and A. Odoni. *Airport Systems Planning, Design, and Management*. New York: McGraw-Hill, 2003.

de Neufville, R., and M. McCord. "Lottery equivalents: Reduction of the certainty effect problem in utility assessment." *Management Science* 32 (1986): 56–60.

de Neufville, R., Y. S. Lee, and S. Scholtes. "Using flexibility to improve value-for-money in hospital infrastructure investments." *Institute of Electrical and Electronic Engineers, Journal of Infrastructure Systems* (2008).

de Neufville, R., S. Scholtes., and T. Wang. "Real options by spreadsheet: parking garage case example." *American Society of Civil Engineers, Journal of Infrastructure Systems* 12 (2) (2006): 107–111.

de Weck, O., R. de Neufville, and M. Chaize. "Staged deployment of communications satellite constellations in low earth orbit." *Journal of Aerospace Computing, Information, and Communication* (2004).

Einstein, A. "On the method of theoretical physics." The Herbert Spencer Lecture, delivered at Oxford (10 June 1933); *Philosophy of Science* 1 (2) (1934): 163–169. http://en.wikiquote.org/wiki/Albert_Einstein.

Feller, W. *An Introduction to Probability Theory and its Applications*. vol. 1. New York: Wiley, 1968.

Fildes, R., and S. Makridakis. "The impact of empirical accuracy studies on time series analysis and forecasting." *International Statistical Review* 63 (3) (1995): 289–308.

Flyvbjerg, B., M. Holm, and S. Buhl. "How (in)accurate are demand forecasts in public works projects? The case of transportation." *Journal of the American Planning Association. American Planning Association* 71 (2) (2005): 131–146.

Flyvbjerg, B., N. Bruzelius, and W. Rothengatter. *Megaprojects and Risk: an Anatomy of Ambition*. New York: Cambridge University Press, 2003.

Forrester, J. *Industrial Dynamics*. Cambridge, MA: MIT Press, 1991.

Franklin, B. Letter to Jean-Baptiste Leroy (13 November, 1789).

Frey, D. D., and H. Wang. Adaptive one-factor-at-a-time experimentation and expected value of improvement. *Technometrics* 48 (2006): 418–431.

Gardner, E. "Exponential smoothing: The state of the art." *Journal of Forecasting* 4 (1985): 1–28.

Gardner, E. "Exponential smoothing: The state of the art—Part II." *International Journal of Forecasting* 22 (2008): 637–666.

Gardner, J. *Excellence*. rev. ed. New York: Norton, 1984.

Gartner, Inc. see Kerner, S. M. *Cell Phones Rising*. Internetnews.com 2005. http://www.internetnews.com/wireless/article.php/3522076

Gessner, G., and J. Jardim. "Bridge within a bridge." *Civil Engineering (New York, N.Y.)* (October) (1998): 44–47.

Gigenrenzer, G. *Reckoning with Risk: Learning to Live with Uncertainty*. London, England: Penguin Books, 2002. Published in the U.S. as *Calculated Risks: How to Know When Numbers Deceive You*. New York: Simon and Schuster, 2002.

González, J., M. Palahí, and T. Pukkala. "Integrating fire risk considerations in forest management in Spain – a landscape level perspective." *Landscape Ecology* 20 (8) (2005): 957–970.

Guma, A. *A Real Options Analysis of a Vertically Expandable Real Estate Development.* Cambridge, MA: S. M. Thesis, MIT Center for Real Estate, 2008.

Guma, A., J. Pearson, K. Wittels, R. de Neufville, and D. Geltner. "Vertical phasing as a corporate real estate strategy and development option." *Journal of Corporate Real Estate* 11 (3) (2009): 144–157.

Hamilton, J. *Time Series Analysis.* Princeton, NJ: Princeton University Press, 1994.

Hassan, R., R. de Neufville, O. de Weck, D. Hastings, and D. McKinnon. "Value-at-risk analysis for real options in complex engineering systems." *Institute of Electrical and Electronic Engineers* (2005).

Hassan, R., and R. de Neufville. "Design of engineering systems under uncertainty via real options and heuristic optimization." Real Options Conference, New York, June (2006).

Innes, J. *Social Indicators and Public Policy: Interactive Processes of Design and Application.* Amsterdam, New York: Elsevier Scientific, 1975.

International Energy Agency (IEA). *World Energy Outlook* (annual).

Jacoby, H., and D. Loucks. "Combined use of optimization and simulation models in river basin planning." *Water Resources Research* 8 (6) (1972): 1401–1414.

Jones, B. "Creech AFB UAV Operations." JonesBlog.

Keeney, R., and R. de Neufville. "Multiattribute preference analysis for transportation systems evaluation: Mexico City airport as a case study." *Transportation Research* 7 (1) (1973): 63–76.

Keeney, R., and H. Raiffa. *Decisions with Multiple Objectives: Preferences and Value Trade-offs.* Cambridge, UK: Cambridge University Press, 1993. Kitco Inc." 5 year copper spot prices" http://www.kitcometals.com/charts/copper_historical_large.html

Kulatilaka, N. "The value of flexibility: The case of a dual-fuel industrial steam boiler." *Financial Management* 22 (3) (1993): 271–280.

Lee, Y. S., and S. Scholtes. "Forecasting for flexible design: Adaptive trend fitting with stochastic forecast errors." Presentation at 2nd International Symposium on Engineering Systems, MIT, June 2009.

Lee, Y. S., and S. Scholtes. *Flexible Capacity Planning for Real Estate Development.* Judge Business School Teaching Case. Cambridge, UK: Cambridge University Press, 2010.

Lee, Y. S., and S. Scholtes. "Empirical prediction intervals and model misspecifications." Working paper, 2010: Judge Business School, University of Cambridge. Cambridge, UK.

Lee, Y. S., R. de Neufville, and S. Scholtes. "Flexibility in hospital infrastructure design." IEEE Conference on Infrastructure Systems, Rotterdam, 10–12 November 2008.

Lieberman, M. "The learning curve and pricing in the chemical processing industry." *Rand Journal of Economics* 15 (1984): 213–228.

Lin, J. *Exploring Flexible Strategies in Engineering Systems Using Screening Models: Applications to Offshore Petroleum Projects.* Cambridge, MA: PhD Dissertation, MIT Engineering Systems Division, 2009.

Lingren, M., and H. Bandhold. *Scenario Planning: The Link Between Future and Strategy.* 2nd ed. New York: Palgrave Macmillan, 2009.

Luenberger, D. *Investment Science.* New York: Oxford University Press, 1996.

Majd, S., and R. Pindyck. "The learning curve and optimal production under uncertainty." *Rand Journal of Economics* 20 (3) (1989): 331–343.

Manne, A., ed. *Investments for Capacity Expansion; Size, Location, and Time-Phasing.* Cambridge, MA: MIT Press, 1967.

Marberry, S., ed. *Improving Health Care With Better Building Design.* Chicago, IL: Health Administration Press, 2006.

Maseda, L. *Real Options Analysis of Flexibility in a Hospital Emergency Department Expansion Project, a Systems Approach.* Cambridge, MA: S.M. Thesis, MIT System Design and Management, 2008.

McConnell, J. *A Life-Cycle Flexibility Framework for Designing, Evaluating, and Managing "Complex" Real Options: Case Studies in Urban Transportation and Aircraft Systems.* Cambridge, MA: PhD Dissertation, MIT Engineering Systems Division, 2007.

Miller, R., and D. Lessard. *The Strategic Management of Large Engineering Projects: Shaping Institutions, Risks, and Governance.* Cambridge, MA: MIT Press, 2001.

Nababan, H. *Dynamic Strategic Planning for Technological Choice: A Case Study of MWRA Residuals Management Facilities Plan.* Cambridge, MA: S.M. Thesis Technology and Policy, 1993.

NAER (Novo Aeroporto SA). "New Lisbon International Airport, Preliminary Planning Study." Final Report, May 1982.

Routledge, R. "The effect of potential catastrophic mortality and other unpredictable events on optimal forest rotation policy." *Forest Science* 26 (3) (1980): 389–399.

Savage, S. *Decision-Making with Insight: Includes Insight.Xla 2.0.* 2nd ed. Pacific Grove, CA: Duxbury Press, 2004.

Savage, S. *The Flaw of Averages.* New York: Wiley, 2009.

Savage, S., S. Scholtes, and D. Zweidler. "Probability management, Part 1." *OR/MS Today* 23 (1) (2006): 21–28.

Savage, S., S. Scholtes, and D. Zweidler. "Probability management, Part 2." *OR/MS Today* 23 (2) (2006): 60–66.

Schoemaker, P. "When and how to use scenario planning: A heuristic approach with illustration." *Journal of Forecasting* 10 (1991): 549–564.

Sharpe, W. "Nuclear financial economics." Research Paper 1275. Stanford University Graduate School of Business, November 1993. http://stanford.edu/~wfsharpe/art/RP1275.pdf.

Shishko, R., R. Aster, and R. Cassingham. *NASA Systems Engineering Handbook.* Washington, DC: National Aeronautics and Space Administration, 1995.

Shumway, R., and D. Stoffer. *Time Series Analysis and Its Applications.* New York: Springer Verlag, 2006.

Sneeuwjagt, R. "Application of wildfire threat analysis in south-western forest of Western Australia." *III International Conference on Forest Fire Research and 14th Conference on Fire and Forest Methodology.* Viegas, D., ed., Portugal *ADAI II*, 1998: 2155–2176.

Steel, K. *Energy System Development in Africa: The Case of Grid and Off-Grid Power in Kenya.* Cambridge, MA: PhD Dissertation, MIT Engineering Systems Division, 2008.

Suh, E. *Flexible Product Platforms.* Cambridge, MA: PhD Dissertation, MIT Engineering Systems Division, 2005.

Teplitz, C. *The Learning Curve Deskbook: A Reference Guide to Theory, Calculations, and Applications.* Westport, CT: Quorum Books, 1991.

Terwiesch, C., and R. Bohn. "Learning and process improvement during production ramp-up." *International Journal of Production Economics* 70 (1) (2001): 1–19.

Thompson, W., I. Vertinsky, H. Schreier, and B. Blackwell. "Using forest fire hazard modeling in multiple use forest management planning." *Forest Ecology and Management* 134 (September 2000): 163–174.

Toffler, A. 1985. *The Adaptive Corporation.* New York: McGraw-Hill.

Tolhurst, K., D. Chong, and N. Stradgard. "Wildfire risk management model for strategic management." *Proceedings of V International Conference on Forest Fire Research.* Viegas, D., ed., Figueira da Foz, Portugal, 27–30 November 2006.

Trigeorgis, L. *Real Options: Managerial Flexibility and Strategy in Resource Allocation. Cambridge.* MA: MIT Press, 1996.

Tromans, N., J. Jefferies, and E. Natamba. "Have women born outside the UK driven the rise in UK births since 2001?" *Population Trends* 136 (Summer 2009): 28–42.

Ulrich, R. "Evidence-based health-care architecture." *Lancet* 368 (2003): S38–S39.

U.S. Energy Information Administration (EIA). (2008) http://www.eia.doe.gov/oiaf/forecasting.html

U.S. Energy Information Administration (EIA). (2010) http://www.eia.doe.gov/emeu/aer/petro.html

U.S. Federal Aviation Administration (FAA). Forecasts and Performance Analysis Division, Office of Aviation Policy and Plans. (2010) *Terminal Area Forecast Summary, Fiscal Years 2009–2030.* http://www.faa.gov/data_research/aviation/taf_reports/media/TAF%20Summary%20Report%20FY%202009%20-%202030.pdf

U.S. Geological Survey. (2010) http://www.usgs.gov/

U.S. Office of Management and Budget. *OMB Circular A-94, Appendix C, Revised Annually.*

Walker, W. *Uncertainty: The Challenge for Policy Analysis in the 21st Century.* Santa Monica, CA: Rand Corporation, 2001.

Wang, T. *Real Options "in" Projects and Systems Design—Identification of Options and Solution for Path Dependency.* Cambridge, MA: PhD Dissertation, MIT Engineering Systems Division, 2005.

Wittels, K., and J. Pearson. *Real Options in Action: Applicability of Vertically Phased Development in Commercial Real Estate.* Cambridge, MA: Joint S.M. Thesis, MIT Center for Real Estate, 2008.

Wooldridge, A. "A survey of telecommunications." *The Economist,* October 9, 1999: 1.

Wright, T. "Factors affecting the cost of airplanes." *Journal of the Aeronautical Sciences* 3 (4) (1936): 122–128.

Yang, Y. *Studying Interdependency Between Multi-level Flexibilities Under Demand Uncertainty.* Cambridge, MA: PhD Dissertation, MIT Engineering Systems Division, 2009.

Zhang, Na. "*Apply Option-Thinking in Long Term Infrastructure Investment: The Case of Commercial Real Estate.*" Cambridge, MA: S.M. Thesis, MIT Technology and Policy Program, 2010.

Index